8

Band 19 · August 1983

isotopenpraxis

AKADEMIE-VERLAG

ISSN 0021–1915 Isotopenpraxis, Berlin 19 (1983) 8, 257–292

EVP 9,– M

An unsere Autoren

Die Zeitschrift veröffentlicht Beiträge zu folgenden Themen: Kontrolle und Analyse radioaktiver und stabiler Isotope, Anwendung radioaktiver und stabiler Isotope in allen Gebieten der Forschung und Technik, der Industrie, Landwirtschaft und Medizin, Herstellung radioaktiver und stabiler Isotope sowie von markierten Verbindungen, kernelektronische Geräte für die Isotopenanwendung (Dickenmesser, Dichtemesser u. ä.) Detektoren für isotopentechnische Arbeiten und den Strahlenschutz in Isotopenlaboratorien, Meßtechnik und Strahlenschutz, Dosimetrie in Isotopenlaboratorien, kerntechnische Geräte für die Isotopenanwendung (Beton, Bestrahlungsanlagen u. ä.), Strahlenchemie, Radiochemie, ökonomische und perspektivische Fragen, Fragen der Weiterbildung.

1. Aufnahme von Manuskripten

Zur Veröffentlichung angenommen werden nur druckreife, noch nicht anderweitig publizierte Beiträge in deutscher, englischer oder russischer Sprache zum Themenbereich der Zeitschrift und zwar

Originalarbeiten
Kurze Mitteilungen
Aus der Praxis
Zusammenfassende Berichte
Tagungsberichte
Informationen.

Sind der Beitrag oder Teile daraus bereits veröffentlicht oder ist eine anderweitige Veröffentlichung vorgesehen, so muß dies der Redaktion mit genauer Angabe der Zweitveröffentlichung mitgeteilt werden. Mit der Annahme des Manuskriptes behält sich der Verlag das Recht der Vervielfältigung, Verbreitung und Übersetzung innerhalb der gesetzlichen Schutzfrist vor. Auszugsweiser Nachdruck oder photomechanische u. a. Wiedergabe sind an eine schriftliche Genehmigung des Verlages gebunden. Manuskripte, Zeichnungen und Photos von veröffentlichten Arbeiten werden nur auf den besonderen und bis zum Zeitpunkt der Veröffentlichung mitgeteilten Wunsch des Autors zurückgesandt.

Bei der unverlangten Einreichung von Manuskripten wird empfohlen, eine Befürwortung eines der Herausgeber oder Redaktionsausschußmitglieder beizufügen.

Manuskripte sind einzureichen bei der Redaktion, 1080 Berlin, Leipziger Str. 3–4

2. Form der Manuskripte

Die Autoren werden gebeten:

2 Manuskripte, eines davon in Original-Maschinenschrift, einseitig und mit 2zeiligem Abstand einzureichen; gut leserliche Photokopien sind zulässig; auf dem 1. Blatt (unter Weglassung der Bezeichnung)

den Titel der Arbeit
Vorname-Initialen und Namen sämtlicher Autoren
gegebenenfalls die Institution, in der sie tätig sind
eine kurze Zusammenfassung der Arbeit von höchstens 15 Zeilen
Keywords in alphabetischer Reihenfolge und
als Fußnote die Anschrift der Institution anzugeben;

wenn möglich, auf gesonderten Blättern Namen, Titel der Arbeit und Zusammenfassung in den beiden anderen obengenannten Sprachen 2fach beizufügen;

Literaturhinweise (fortlaufend numeriert), Fußnoten, Tabellen, Abbildungen (in 2–3facher Veröffentlichungsgröße), Abbildungsunterschriften (getrennt von den Abbildungen mit Zeichenerklärungen, Quellenangaben, falls erforderlich, und gegebenenfalls mit einem Vermerk für besondere Wünsche bezüglich der Abbildungsgröße), private und Dienstanschrift der Verfasser mit akademischen Titeln und ausgeschriebenen Vornamen auf besonderen Blättern ans Ende des Manuskriptes zu stellen;

Wünsche für den Kolumnentitel (Seitenüberschrift) bekanntzugeben;

Absätze durch Einzüge von mindestens 3 Anschlägen kenntlich zu machen;

die Maßeinheiten des SI-Systems ("Systeme International") zu verwenden;

Korrekturen im Manuskript unmittelbar an der zu korrigierenden Stelle anzubringen;

Formeln deutlich lesbar und möglichst groß einzutragen;

Farbabbildungen und nachträgliche Änderungen oder Ergänzungen in Abbildungen zu vermeiden;

hervorzuhebende Textstellen zu unterstreichen, und zwar halbfetten Satz mit einer geraden schwarzen Linie, kursiven Satz (Schrägschrift) mit einer blauen Wellenlinie (alle Eigennamen). Sperrungen gestrichelt;

besondere Buchstaben durch farbige Unterstreichung zu kennzeichnen, und zwar Formelzeichen in

Griechisch-Kursiv (α, β, ...) rot,
Kursiv (A, B, ...) blau
halbfetter Kursiv ($\boldsymbol{A}$, $\boldsymbol{B}$, ...) 2 Linien blau (Vektoren),
halbfetter Grotesk (**A**, **B**, ...) 2 Linien violett (Tensoren);

bei Literaturangaben auf Vollständigkeit zu achten. In nachstehender Reihenfolge müssen angegeben sein:

a) bei *Zeitschriften*: Vorname-Initialen und Namen sämtlicher Autoren, Titel der Zeitschrift (nach „Periodica Chimica" oder möglichst wenig gekürzt), Bandnummer (unterstrichen), Jahreszahl (in Klammern), Heftnummer (nur bei Zeitschriften, die jedes Heft mit S. 1 beginnen) sowie die Seitenzahl, auf der die zitierte Arbeit beginnt.

b) Bei *Büchern*: Initialen der Vornamen und Name des Autors (Herausgeber werden durch den Zusatz „ed." gekennzeichnet), voll ausgeschriebener Titel des Buches in der Originalsprache, evtl. Bandnummer, voll ausgeschriebener Name des Verlages, Nummer der Auflage (außer der ersten), Erscheinungsort(e), Erscheinungsjahr (hier nicht in Klammern setzen!) und evtl. Kapitelnummer, Seitenzahl u. ä.

c) bei *Sammelbänden, Tagungsberichten*: Es soll zusätzlich der Titel des zitierten Beitrags angegeben werden. Die Seitenzahl muß unbedingt angegeben werden. Aus dem Zitat muß hervorgehen, ob es sich um einen Tagungssammelband oder um einen noch nicht veröffentlichten Vortrag handelt. Zwischen Tagungsort und Erscheinungsort des entsprechenden Tagungsberichts ist zu unterscheiden.

3. Korrekturen

Der Autor erhält zwei Umbruchabzüge, von denen einer möglichst schnell korrigiert an die Redaktion zurückgereicht werden soll. Auf Wunsch des Autors kann die Korrektur auch von der Redaktion gelesen werden.

Durch nachträgliche Manuskriptänderungen verursachte Korrekturkosten gehen, sofern sie 10% der Satzkosten übersteigen, zu Lasten des Autors.

Der Autor erhält 50 Sonderdrucke kostenlos. Weitere Sonderdrucke können bis zur Rückgabe der Korrekturabzüge auf Kosten des Autors bei der Redaktion bestellt werden.

Bezugsmöglichkeiten

Bestellungen sind zu richten

- **in der DDR** an den Postzeitungsvertrieb unter Angabe der Kundennummer des Bestellenden oder an den AKADEMIE-VERLAG, DDR-1086 Berlin, Leipziger Straße 3–4, PF-Nr. 1233;
- **im sozialistischen Ausland** an eine Buchhandlung für fremdsprachige Literatur oder an den zuständigen Postzeitungsvertrieb;
- **in der BRD und Berlin(West)** an eine Buchhandlung oder an die Auslieferungsstelle KUNST UND WISSEN, Erich Bieber OHG, Wilhelmstraße 4–6, D-7000 Stuttgart 1;
- **in den übrigen westeuropäischen Ländern** an eine Buchhandlung oder an die Auslieferungsstelle KUNST UND WISSEN, Erich Bieber GmbH, Dufourstraße 51, CH-8008 Zürich;
- **im übrigen Ausland** an den Internationalen Buch- und Zeitschriftenhandel; den Buchexport, Volkseigener Außenhandelsbetrieb der Deutschen Demokratischen Republik, DDR-7010 Leipzig, Postfach 160; oder an den AKADEMIE-VERLAG, DDR-1086 Berlin, Leipziger Straße 3–4.

Zeitschrift „Isotopenpraxis"

Herausgeber: Im Auftrag des Akademie-Verlages herausgegeben von Prof. Dr. Justus Mühlenpfordt
Verlag: Akademie-Verlag, DDR-1086 Berlin, Leipziger Straße 3–4, PF-Nr. 1233; Fernruf: 2236229; Telex-Nr.: 114420;
Bank: Staatsbank der DDR, Berlin, Kto.-Nr.: 6836-26-20712
Verantwortlich für den Inhalt des Heftes: Prof. Dr. H. Hübner, Leipzig
Anschrift der Redaktion (Fernruf Frau H. Schier: 6351177) der Zeitschrift Isotopenpraxis: Akademie-Verlag, DDR-1086 Berlin, Leipziger Str. 3–4, PF-Nr. 1233
Veröffentlicht unter der Lizenznummer 1527 des Presseamtes beim Vorsitzenden des Ministerrates der Deutschen Demokratischen Republik
Gesamtherstellung: VEB Druckerei „Thomas Müntzer", DDR-5820 Bad Langensalza
Erscheinungsweise: Die Zeitschrift „Isotopenpraxis" erscheint monatlich. Die 12 Hefte eines Jahrganges bilden einen Band. Bezugspreis je Band 180,– M zuzüglich Versandspesen Preis für die DDR 108,– M) Preis je Heft 15,– (Preis für die DDR 9,– M)
Bestellnummer dieses Heftes: 1079/19/8

AN (EDV) 12831

ISOTOPENPRAXIS

Band 19 (1983) Heft 8, S. 257—292

Redaktion und Verlag: Akademie-Verlag, Redaktion Isotopenpraxis, 1086 Berlin, Leipziger Straße 3—4

Inhaltsverzeichnis

Originalarbeiten

K. Gloe, P. Mühl

Bestimmung von Metallextraktionsparametern mit Hilfe der Tracertechnik (*in Englisch*)

Isotopenpraxis **19** (1983) H. 8, S. 257 bis 260
10 Abb., – Tab., 9 Lit.

Die Flüssig-flüssig-Extraktion ist eine sehr effektive Trenn- und Anreicherungsmethode für Metallionen aus der wäßrigen Phase. Es werden Experimente beschrieben, die insbesondere eine automatische Ermittlung von Prozeßparametern unter Verwendung von Radiotracern erlauben. Aus den Ergebnissen können verschiedene Einflußfaktoren auf die Wirksamkeit des angewendeten Extraktionssystems erhalten werden. Einige Beispiele werden vorgestellt.

T. Oi, Y. Sakuma, M. Maeda, M. Okamoto

Trennung von Uranisotopen bei der Kationenaustausch-Chromatographie des Systems Uran(IV)—Zitronensäurekomplexe (*in Englisch*)

Isotopenpraxis **19** (1983) H. 8, S. 261 bis 264
6 Abb., 2 Tab., 9 Lit.

Die Trennung der Uranisotope ^{235}U und ^{238}U, die bei der Kationenaustausch-Chromatographie eines Systems Uran(IV)—Zitronensäurekomplex, das Schwefelsäure enthält, vor sich geht, wurde bei verschiedenen Temperaturen untersucht. Ein eindeutiges Umlagerungschromatogramm wurde nur innerhalb einer begrenzten Zusammensetzung und eines begrenzten pH-Wertes des Elutionsmittels erzielt. Es wurde festgestellt, daß an der Uranbandenfront eine Verarmung an ^{235}U eintritt, während der hintere Bereich angereichert wird. Dieses Ergebnis ist dem bei U(IV)-Komplexbildungssystemen ermittelten entgegengesetzt. Der Einstufen-Trennfaktor ε für $^{235}U/^{238}U$ (definiert durch Gleichung (4)), sank mit steigender Temperatur.

R. Otto, P. Hecht

Zur Markierung mit Generatornukliden (*in Englisch*)

Isotopenpraxis **19** (1983) H. 8, S. 265 bis 267
4 Abb., 3 Tab., 1 Lit.

Kurzlebige Radionuklide aus Radionuklidgeneratoren werden in der letzten Zeit zunehmend bei industriellen Tracerexperimenten genutzt, und häufig sind organische Phasen und auch feste Stoffe zu markieren. Es wird deshalb über Untersuchungen zur Überführung der Tochternuklide der handelsüblichen Mo/Tc-, Sn/In- und Te/I-Generatoren und eines selbstaufgebauten Ba/La-Generators aus den Generatoreluaten in organische Phasen und über die Markierung öliger und fester Stoffe berichtet. ^{99m}Tc, ^{113m}In und ^{132}I werden z. B. durch Extraktion in organische Lösungsmittel übergeführt. Nach einer Extraktion des ^{132}I mit einer ethanolischen KI/I_2-Lösung lassen sich durch Isotopenaustausch jodorganische Verbindungen herstellen. Zur Markierung von Ölen eignen sich ^{113m}In oder ^{140}La-oleate. Bei der Markierung fester Phasen werden die adsorptiven Eigenschaften der Feststoffe genutzt. Abschließend werden einige praktische Anwendungsbeispiele besprochen.

I. Pavlicsek, N. Stenger, A. Veres

Gerät für die Gamma-Aktivierungsanalyse (*in Englisch*)

Isotopenpraxis **19** (1983) H. 8, S. 268 bis 269
2 Abb., 2 Tab., 5 Lit.

Eine Bestrahlungsanlage — 740 TBq (20,000 Ci) ^{60}Co mit einer Intensität von $7{,}74 \cdot 10^3$ C/kg (30 MR/h) — sowie eine pneumatische Rohrpostanlage die von einem mit geeignetem Programm versehenen automatischen System gesteuert wird, wurde gebaut. Der Gehalt an Se, Br, Sr, Ag, Cd, I, Hf, Ir, Au und Pt kann bei Proben verschiedener Herkunft in einem Volumen von 5 cm³ mit einer Genauigkeit von 0,05—1 mg/g bestimmt werden.

Contents

Articles

K. Gloe, P. Muehl

Determination of Metal Extraction Process Parameters Using Tracer Technique (*in Englisch*)

Isotopenpraxis **19** (1983) No. 8, p. 257 to 260
10 Fig., – Tab., 9 Ref.

Liquid-liquid-extraction is an effective separation and concentration process for metal ions in aqueous solution. The paper shows the experimental technique of extraction experiments, especially the automatical measurement of process parameters. The use of radioactive tracers during these investigations allows to obtain basic information about the influence of numerous factors in an extraction system. Examples are discussed.

T. Oi, Y. Sakuma, M. Maeda, M. Okamoto

Isotope Fractionation of Uranium in Cation Exchange Chromatography of U(IV)—Citric Acid Complexes (*in English*)

Isotopenpraxis **19** (1983) No. 8, p. 261 to 264
6 Fig., 2 Tab., 9 Ref.

The $^{235}U/^{238}U$ isotope fractionation occuring in cation exchange chromatography of a system U(IV)—citric acid containing sulfuric acid was investigated at various temperatures. A clear displacement chromatogram was obtained at only a limited composition and pH of the eluent. It was found that ^{235}U is depleted at the front of the uranium band, and enriched at the rear. This result was opposite to that observed in the U(VI) complexation systems. The $^{235}U/^{238}U$ single stage separation coefficient ε (defined by Eq. (4)) decreased with rising temperature.

R. Otto, P. Hecht

On Labelling with Generator Nuclides (*in English*)

Isotopenpraxis **19** (1983) No. 8, p. 265 to 267
4 Fig., 3 Tab., 1 Ref.

Short-lived radioisotopes separated from radionuclide generators are widely used. In the paper possibilities of labelling with the daughter nuclides of the commercially available Mo/Tc-, Sn/In- and Te/I-generators and of the self-made Ba/La-generator in industrial tracer experiments are presented. The transfer of the daughter nuclides from the generator eluates into organic phases and the labelling of oil-phases and solid particles are investigated. The developed simple, quick and efficient methods are suitable for routine application under industrial conditions. Some examples of industrial applications of the generator nuclides are given, too.

I. Pavlicsek, V. Stenger, A. Veres

Apparatus for Gamma Activation Analysis (*in English*)

Isotopenpraxis **19** (1983) No. 8, p. 268 to 269
2 Fig., 2 Tab., 5 Ref.

An irradiation device containing 740 TBq (20,000 Ci) ^{60}Co giving $7{\cdot}74 \times 10^3$ C/kg intensity (30 MR/h) and a pneumatic dispatch rabbit system controlled by an automatic system provided with an appropriate programme was built. The Se, Br, Sr, Ag, Cd, I, Hf, Ir, Au and Pt content of samples of different origin can be determined in a volume of 5 cm³ with 0,05—1 mg/g sensitivity.

P. Urbański, M. Jankowska, E. Kowalska, D. Wagner

Bestimmung von Ca und Fe und Messung des Aschegehalts in Braunkohle (*in Englisch*)

Isotopenpraxis **19** (1983) H. 8, S. 270 bis 272
4 Abb., – Tab., 5 Lit.

Ein Versuch zur Bestimmung von Kalzium und Eisen sowie zur Messung des Aschegehalts in Braunkohle wurde mit Hilfe des XRF-Verfahrens und der niedrigenergetischen Röntgenrückstreuung unternommen. Verwendet wurden eine ^{238}Pu-Quelle, ein argongefülltes Proportionalzählrohr und ein Dreikanal-Impulshöhenanalysator. Ein einfaches theoretisches Modell wird vorgestellt; die erzielten Ergebnisse werden diskutiert.

P. Urbański, M. Jankowska, E. Kowalska, D. Wagner

Determination of Ca and Fe and Measurement of Ash Content in the Brown Coal (*in English*)

Isotopenpraxis **19** (1983) No. 8, p. 270 to 272
4 Fig., – Tab., 5 Ref.

Applying the XRF technique and the low-energy X-ray backscattering method an attempt of calcium and iron determination, as well as measurement of the ash content in the brown coal was made. A ^{238}Pu source, an argon filled proportional counter and a three channel pulse height analyser were used. A simple theoretical model is proposed and obtained results are discussed.

Kurze Originalmitteilungen

H.-G. Könnecke, D. Luther

Einfluß von Belüftungssystemen auf die Verweilzeit in biologischen Kläranlagen (*in Englisch*)

Isotopenpraxis **19** (1983) H. 8, S. 273 bis 274
4 Abb., – Tab., 3 Lit.

Y. M. Megahed, M. F. Abdel-Wahab, M. A. Zewail, A. M. Shalaby

Festphasen-Peptidsynthese von mit ^{125}I markierten Insulin-Fragmenten. Teil 1: Synthese von mit ^{125}I markiertem Insulinfragment B_{3-6}-Asparagin—Glutamin—Histidin—Leucin—OCH_3 (*in Englisch*)

Isotopenpraxis **19** (1983) H. 8, S. 274 bis 276
4 Abb., 1 Tab., 12 Lit.

Informationen

Isotopenpraxis **19** (1983) H. 8, S. 276 bis 282

Buchbesprechungen

Isotopenpraxis **19** (1983) H. 8, S. 282 bis 292

Letters to the Editor

H.-G. Koennecke, D. Luther

The Influence of Airing Systems on the Residence Time in Biological Clarification Plants (*in English*)

Isotopenpraxis **19** (1983) No. 8, p. 273 to 274
4 Fig., – Tab., 3 Ref.

Y. M. Megahed, M. F. Abdel-Wahab, M. A. Zewail, A. M. Shalaby

Solid Phase Peptid Synthesis of ^{125}I Labelled Insulin Fragments. Part 1: Synthesis of ^{125}I Labelled Insulin Fragment B_{3-6}-Asparagine—Glutamine—Histidine—Leucine—OCH_3 (*in English*)

Isotopenpraxis **19** (1983) No. 8, p. 274 to 276
4 Fig., 1 Tab., 12 Ref.

Information

Isotopenpraxis **19** (1983) No. 8, p. 276 to 282

New Books

Isotopenpraxis **19** (1983) No. 8, p. 282 to 292

Isotopenpraxis 19 (1983) 8, S. 257–260

Determination of Metal Extraction Process Parameters Using Tracer Technique*)

K. Gloe, P. Mühl

(Central Institute of Solid State Physics and Material Research,
Academy of Science of the GDR, Dresden)[1])

Liquid-liquid-extraction is an effective separation and concentration process for metal ions in aqueous solution. The paper shows the experimental technique of extraction experiments, especially the automatical measurement of process parameters. The use of radioactive tracers during these investigations allows to obtain basic information about the influence of numerous factors in an extraction system. Examples are discussed.

Keywords

distribution functions; extraction apparatuses; hydrometallurgy; metals; solvent extraction; tracer techniques

In the recent years hydrometallurgical processes of metal production became more and more importance [1, 2]. In connection with this tendency the liquid-liquid-extraction as technological step was progressively of high interest. This development is based both on economical and technical conditions. Extraction processes permit the metal winning from low-grade and complex ores and secondary sources with high separation selectivity and effectivity. They offer advantageous suppositions for continuous and energy concious operating. In the field of reprocessing of nuclear fuel they permit high separation and decontamination factors under extreme working conditions [3–5]. The general block scheme of an extraction process, as it is used in large scale for instance for the copper winning, is shown in Fig. 1. At the extraction section the transition of the metal ion into the organic phase takes place by contacting the aqueous metal salt solution with the organic phase containing the complexing agent, the extractant. After a possible scrub step in order to increase the separation effectivity in relation to contamination metals the metal is stripped into a new pure aqueous phase, from which it can be recovered by usual methods. The metal-free organic phase is recycled to the extraction section.

The liquid-liquid extraction technique has a wide variety of application possibilities caused by the numerous parameters of influence responsible for the phase transition of the metal ions. On the other hand a high experimental and theoretical effort must be done to obtain optimal conditions. In Fig. 2 a step-by-step development of an extraction process is given. It is shown that the selection of an extraction system and of the laboratory scale testing are influenced by a lot of several factors, which are necessary for the technical application. The distribution ratio, D_M, is the decisive parameter, which characterizes the mass transfer between aqueous and organic phase in dependence on the mentioned parameters of influence. D_M represents the quotient of the analytical concentrations of the metal in both phases ($D_M = c_{M(org)}/c_{M(w)}$).

For the determination of the metal concentration the application of radioactive tracers is preferable. This method of determination of the concentration has the advantage that the concentrations can be obtained independently in both phases and without any additional separating operations. The sensitivity can be varied in wide ranges according to the requires by choice of the specific activity. Furthermore it is possible getting information about the distribution behaviour of extractants and diluents with the ^{14}C labelled reagents. Such data are necessary for the quantitative treatment of extraction reactions.

Traditionaly extraction investigations are performed in separating funnels. An apparatus of this kind is shown in

*) Presented at the 2nd Working Meeting "Radioisotope Application and Radiation Processing in Industry", 28th September to 1st October 1982, Leipzig

1) Address: DDR-8027 Dresden, Helmholtzstr. 20

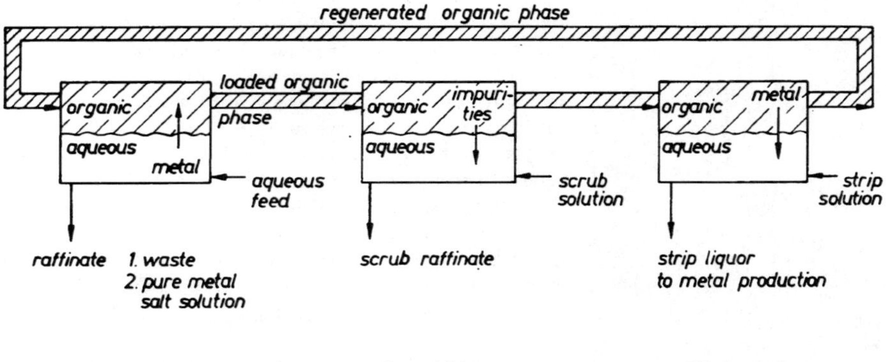

Fig. 1. Scheme of an extraction process

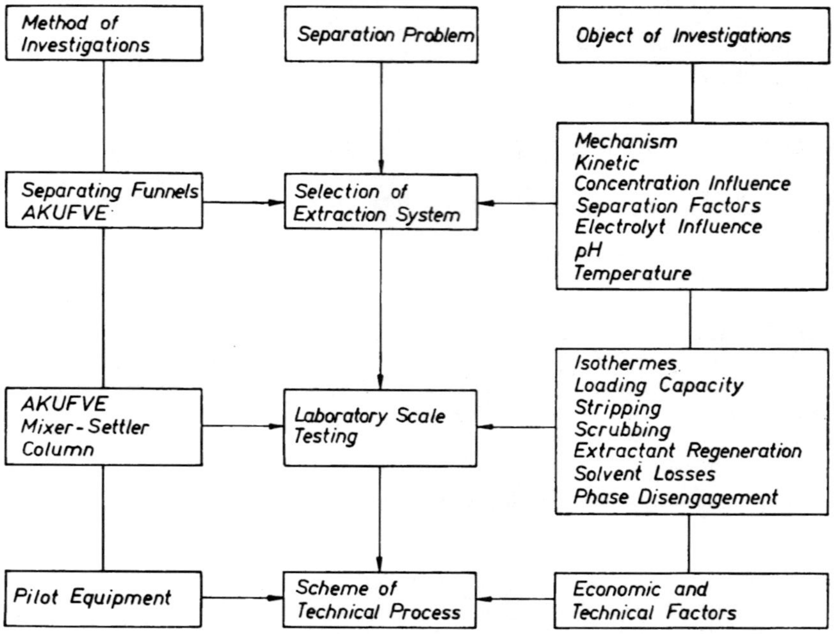

Fig. 2. Step-by-step development of an extraction process

Fig. 3. Shaking apparatus with separating funnels

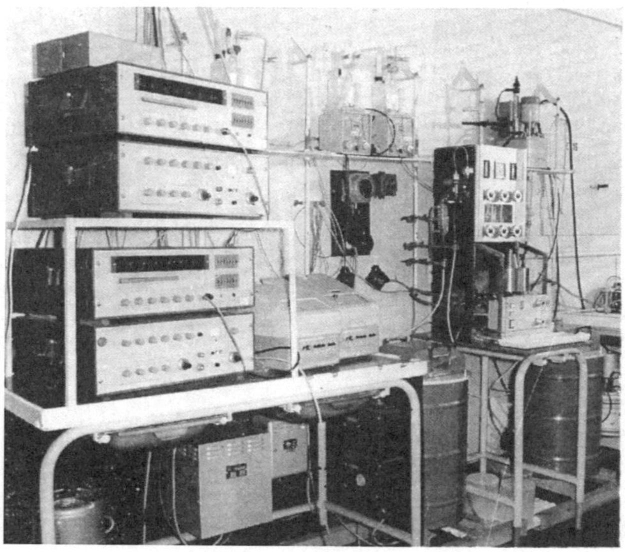

Fig. 4. Extraction apparatus AKUFVE in laboratory operation

Fig. 3. Thereby for each experimental value a separate test is necessary. The demand will increase with variation of the parameters of influence given in Fig. 2 and moreover to a high degree it depends on the operator's skill. Certain investigations, for instance at constant *p*H-value or constant distribution ratio D_M, which are very important for the determination of special reaction parameters, either cannot at all or can be performed, but only with a great number of experiments. An essential development is represented by the application of the automatic extraction apparatus AKUFVE [6]. Fig. 4 demonstrates the AKUFVE-apparatus in laboratory operation. Such an equipment allows to measure the distribution ratios continuously varying the parameters of interest. Both phases are mixed intensively in a mixing chamber. Then they are separated in an ultracentrifuge, analyzed with different detectors — in our case the γ-activity is measured by means of a scintillation counting system — and the solutions are drived back to the mixing chamber [7]. The developed processing schemes were tested on laboratory scale. The equipment in the authors' laboratory is a mixer-settler-apparatus with 12 theoretical stages operating continuously.

By means of the presented experimental methods the extraction processes given in Fig. 5 have been developed and tested in laboratory scale. The radionuclides used during these investigations are mentioned there too. In the following the course of development of two processes shall be interpreted. At the production of alumina from aluminosilicate ores by means of acid leaching it is necessary to separate iron, which is dissolved together with aluminium. One of the possibilities is the extractive separation with aliphatic monocarboxylic acids as extractant. Fundamental investigations on the mechanism has shown that iron(III) is extracted as an trimeric complex $(FeR_3 \times H_2O)_3$ [8].

In Fig. 6 the log D_{Fe}-*p*H-diagram for the extraction of iron from sulfate solution is given. The different slope as well as the dependence of the distribution ratio on the iron concentration point to the polymerization of the formed complexes in the organic phase. At the technical applica-

Extraction	Extractant	Radiotracer
Nb/Ta-separation	Methyl isobutyl ketone	^{95}Nb, ^{182}Ta, ^{187}W, ^{54}Mn, ^{51}Cr
Mo/W-separation	Tributyl phosphate	^{99}Mo, ^{187}W, ^{60}Co, ^{59}Fe
Fe/Al-separation	Carboxylic acid, Tributyl phosphate, Amine	^{59}Fe, ^{14}C, ^{32}P
$FeCl_3$-purification	Carboxylic acid, Tributyl phosphate	^{59}Fe, ^{60}Co, ^{65}Ni, ^{51}Cr, ^{54}Mn, ^{64}Cu
$ZnCl_2$-purification	Alkyl phosphoric acid, Acylthiourea	^{65}Zn, ^{59}Fe, ^{64}Cu, ^{65}Ni, ^{51}Cr, ^{14}C
$Pb(NO_3)_2$-purification	Carboxylic acid	^{110m}Ag, ^{64}Cu, ^{59}Fe
$CuCl_2$-purification	Acylthiourea	^{64}Cu, ^{59}Fe, ^{60}Co, ^{65}Ni, ^{65}Zn, ^{14}C
V_2O_5-purification	Carboxylic acid	^{59}Fe, ^{60}Co, ^{65}Ni, ^{51}Cr
Hg-recovery	Acylthiourea	^{203}Hg, ^{14}C

Fig. 5. Developed extraction processes and used radionuclides

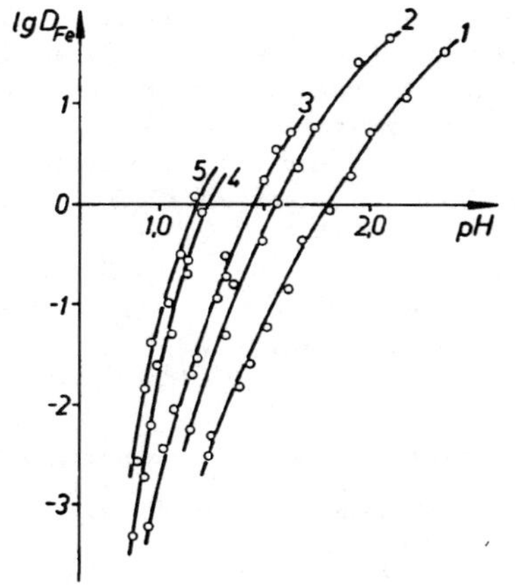

Fig. 6. Iron(III) extraction with n-caprylic acid in dependence on the iron concentration and pH

$c_{Fe(total)}$ 1 — 5×10^{-4} M; 2 — 1×10^{-3} M; 3 — 4×10^{-3} M; 4 — 5×10^{-2} M; 5 — 1×10^{-1} M; $c_{HR(org)} = 6{\cdot}3$ M

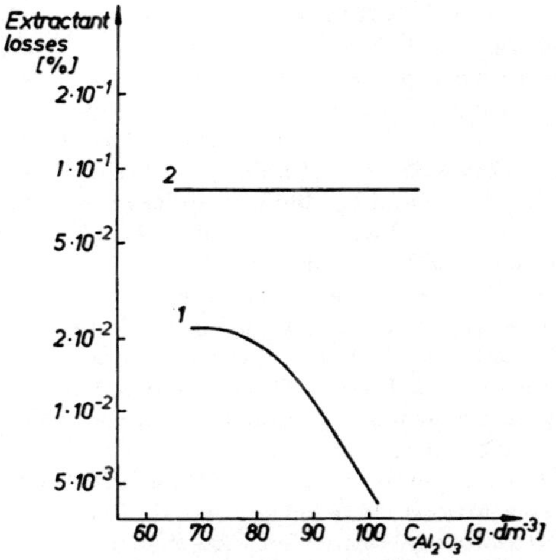

Fig. 8. Extractant losses of n-caprylic acid in dependence on the aluminium concentration

1 — 80 % n-caprylic acid/20 % decane; 2 — 100 % n-caprylic acid

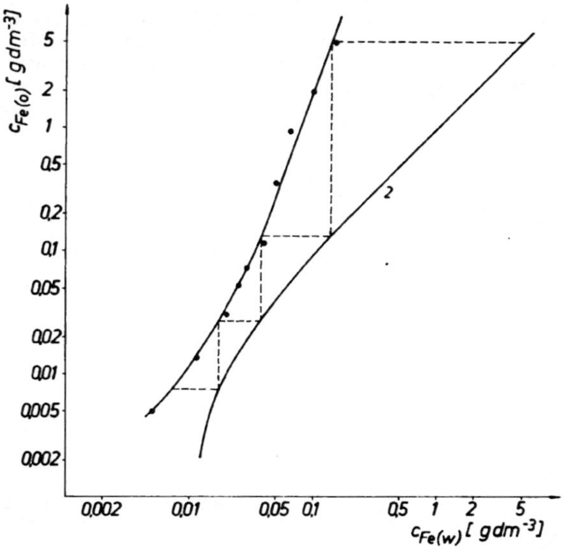

Fig. 7. Isotherme of the iron extraction from concentrated $Al_2(SO_4)_3$ solution

1 — equilibrium line $c_{Al_2(SO_4)_3} = 280$ g $\times$ dm^{-3}; c_{HR} = 80 % PC-acid/20 % Paraffin I;
2 — operating line $pH = 2{\cdot}5$

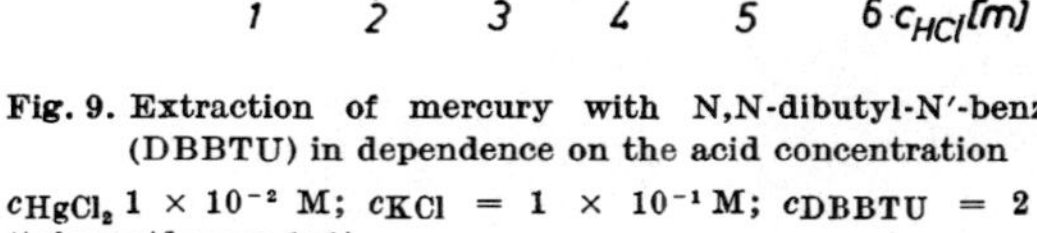

Fig. 9. Extraction of mercury with N,N-dibutyl-N′-benzoylthiourea (DBBTU) in dependence on the acid concentration

c_{HgCl_2} 1×10^{-2} M; $c_{KCl} = 1 \times 10^{-1}$ M; $c_{DBBTU} = 2 \times 10^{-1}$ M (toluene/decane 1:1)

tion this result is very important. Considering the isotherme of the iron(III) extraction from aluminium sulfate solution, shown in Fig. 7, it is obvious that the demand for the separation of low quantities of iron considerably increases owing to the polymerization. The needed number of theoretical stages for the reduction of the iron content from nearly 5 g · dm^{-3} down to $\leq 0{\cdot}01$ g · dm^{-3} is determined to

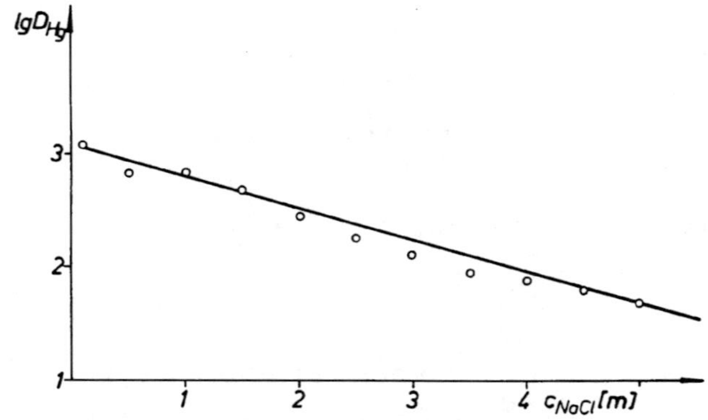

Fig. 10. Extraction of mercury with N,N-dibutyl-N'-benzoylthiourea in dependence on the sodium chloride concentration
$c_{HgCl_2} = 1 \times 10^{-2}$ M; $c_{NaCl} = 0{,}1-5$ M; $c_{DBBTU} = 2 \times 10^{-1}$ M (toluene/decane 1:1)

4 according to McCabe/Thiele, which could be confirmed in the laboratory test [9].

For the economy of the process the losses of extractants are very decisive. These are losses caused by solubility, evaporation, decomposition and insufficient phase disengagement. Investigations to determine the extractant losses caused by solubility have been carried out using n-caprylic acid labelled with ^{14}C. In Fig. 8 these losses of the carboxylic acid are given in dependence on the content of aluminium in the aqueous initial solution. It is obvious that the losses are relatively high using undiluted carboxylic acid as extractant, but they can be decreased considerably by addition of a diluent. The extractant losses are decreased by an order of magnitude and more in case of the addition of 20% decane as diluent.

As second example the separation of mercury from aqueous process solutions shall be illustrated. Because of the enormous toxicity and connected with this of the necessity of an enviromental concious processing as well as of the economical importance to recover mercury from industrial waste water the development of a mercury removal process is especially burning. Furthermore the attainable extremely low residual concentration of mercury is a decisive demand. The liquid-liquid extraction of mercury using substituted acylthioureas of the general formula $RCONHCSNR_2$ permits the separation of mercury in a wide range of acid concentrations. In Fig. 9 the influence of the hydrochloric acid concentration on the extraction of mercury is shown. The distribution ratio D_{Hg} is nearly constant in a neutral and weak acid range. A slight decrease of distribution ratios can be observed only at higher acid concentrations. In the same order the distribution ratios D_{Hg} decreases with rising sodium chloride concentration as shown in Fig. 10. Even at a concentration of 5 M the distribution ratios are >50 (that means the extraction is $>98\%$), so that the separation of mercury is always guaranteed. It is possible to concentrate both in the extraction section and during stripping by changing the volume ratio between aqueous and organic phase.

Summarizing it can be stated that for the development of liquid-liquid extraction processes the application of tracer technique is the preferred method in order to determine fundamental reaction parameters as well as technological parameters. As is well known the method provides a high detection sensitivity and is time-saving, especially also at continuously operating equipments. Generally the influence on the results by variation of processing conditions, like concentrations, *p*H-value, temperature and so on, is low. Another advantage is given by the possibility of labelling of the main and contamination metals and of the subsequent determination of their distribution behaviour. Moreover it is possible to determine the distribution of extractants and diluents in a simple way when the synthesis of ^{14}C-labelled compounds is realizable.

Received January 14, 1983

References

[1] *J. A. Holmes*, Paper of the symposium "Hydrometallurgy 81", Soc. Chem. Ind., Manchester 30. 6. – 3. 7. 1981, A1, S. 1 – 14
[2] *P. Mühl, K. Gloe*, Mitt.-Bl. Chem. Ges. DDR **27** (1980) 153
[3] *D. S. Flett*, Chem. Ind. **1977**, 706
[4] *H. Reinhardt*, Paper of the symposium "Hydrometallurgy 81", Soc. Chem. Ind., Manchester 30. 6. – 3. 7. 1981, F1, S. 1 – 11
[5] *A. Chesné*, Paper of the ISEC 80, Liege 6. – 12. 9. 1980, Vol. 1, 80 – III, S. 1 – 9
[6] *J. Rydberg, H. Reinhardt, J. O. Liljenzin*, in: Ion Exchange and Solvent Extraction, ed. by *J. A. Marinsky* and *Y. Marcus*, Marcel Dekker Inc., New York, 1973, Vol. 3, p. 111
[7] *K. Gloe, P. Mühl*, Isotopenpraxis **15** (1979) 236
[8] *P. Mühl, A. I. Kholkin, L. M. Gindin, K. Gloe, K. S. Luboshnikova*, Proc. ISEC 74, Lyon, Soc. Chem. Ind., London **1974**, Vol. 2, p. 1717
[9] *P. Mühl, K. Gloe, C. Fischer, S. Ziegenbalg, H. Hoffmann*, Hydrometallurgy **5** (1980) 161

Isotopenpraxis 19 (1983) 8, S. 261–264

Isotope Fractionation of Uranium in Cation Exchange Chromatography of Uranium(IV) — Citric Acid Complexes

T. Oi[1]), *Y. Sakuma*[1]), *M. Maeda*[2]), *M. Okamoto*[1])

(Research Laboratory for Nuclear Reactors, Tokyo Institute of Technology[1]) and Department of Applied Chemistry, Nagoya Institute of Technology-))

The $^{235}U/^{238}U$ isotope fractionation occuring in cation exchange chromatography of a system U(IV)—citric acid containing sulfuric acid was investigated at various temperatures. A clear displacement chromatogram was obtained at only a limited composition and pH of the eluent. It was found that ^{235}U is depleted at the front of the uranium band, and enriched at the rear. This result was opposite to that observed in the U(VI) complexation systems. The $^{235}U/^{238}U$ single stage separation coefficient ε (defined by Eq. (4)) decreased with rising temperature.

Keywords

citric acid; ion exchange chromatography; isotope separation; separation processes; uranium ions; uranium isotopes

Introduction

Ion exchange chromatography for uranium isotope fractionation has been applied successfully to the systems containing uranium(VI) ions(uranyl ions). In a previous paper [1], we reported the results of uranium isotope fractionation in cation exchange chromatography of uranyl-carboxylate and -fluoride complexes. It was found that the lighter isotope ^{235}U is accumulated preferentially at the frontal part of the chromatogram and ^{238}U at the rear part, and that the former isotope is more enriched in the complex than in the aqua uranyl ion, the magnitude being in the order acetate < fluoride < citrate. These findings indicate that the larger the stability constant of the complex, the more is ^{235}U enriched in the complex. Reduced partition function ratios for uranium isotopes were calculated on the basis of the Raman and infrared spectra of the O=U=O bond frequencies of the complexes in aqueous solution, and it was suggested that the reduction of the O=U=O bond strength by complex formation may be a predominant factor for the above observations [2, 3].

On the other hand, since the uranium(IV) ions(uranous ions) is nonoxygenated unlike the uranyl ion, complex formation equilibria with ligands and resin phases should be different from those of the uranyl ion, and consequently different types of isotope effects may be expected. Few investigations have been carried out on ion exchange chromatography of uranous complexes, probably because the uranous ion is more strongly adsorbed on cation exchange resin and more susceptible to hydrolysis than the uranyl ion, and is oxidized to the uranyl ion. In a recent paper [4], we described an ion exchange chromatographic process for uranous complexes. Uranous ions were loaded onto a cation exchange resin column in the H^+ form to prepare a band of uranous ions, and then the band was eluted with solution containing both magnesium sulfate and sulfuric acid. It was made clear that ^{235}U is depleted at the front of the band and enriched at the rear. This trend is contrary to that observed in the uranyl complex formation systems [1].

Addresses: [1]) Ookayama, Meguro-ku, Tokyo 152, Japan; [2]) Gokiso-cho, Showa-ku, Nagoya 466, Japan

In the present work, in order to gather basic data concerning uranium isotope fractionation in systems involving uranous complex formation, we carried out a cation exchange band elution process for a system uranous — citric acid, to which sulfric acid and potassium sulfate were added in order to reduce the oxidation rate of uranous ions and to lower the *p*H of the eluent.

Experimental

Reagents. The ion exchange resin was a high-porous-type strongly acidic cation exchange resin PK-1 (100—200 mesh) supplied from Asahi Chemical Industry. The stock solution of uranous ions was prepared by dissolving natural uranium metal (atomic ratio $^{235}U/^{238}U = 0{\cdot}007280 \pm 0{\cdot}00004$) in concentrated hydrochloric acid and filtering off the residue. All other reagents used were of analytical grade.

Columns. The columns used were of the same type as that illustrated in Ref. [4]. They were of pressure-resistant Pyrex glass, and had 1 cm inner diameter and 110 cm resin bed length. The column was surrounded by a jacket through which thermostated water was circulated. In chromatographic experiments three columns were connected in series by Teflon tubes having an inner diameter of 1 mm.

Chromatographic procedures. The cation exchange resin was packed in the columns, and pre-treated with 2 mol dm^{-3} H_2SO_4 and 1 mol dm^{-3} HCl to convert to H^+ form and then washed thoroughly with distilled water. A uranous load solution was fed into the first column to prepare a uranous band of a suitable length (20—50 cm). Chromatography was then carried out by feeding an eluent continuously to the column using a constant flow-rate peristaltic pump. The displacement rate of the band was kept at about 1 cm/h. The effluent from the bottom of the third column was collected in fractions of about 1·5 ml. The *p*H of the effluent was measured with a *p*H meter at several fraction intervals. The concentration of the uranous ion in each fraction was determined by a colorimetric method and the isotopic analysis of uranium was carried out using a Varian Mat CH 5 mass spectrometer [1]. Prior to the isotopic analysis sample fractions were pre-treated as follows. Uranous ions

Tab. 1. Experimental conditions and results in chromatography at 50 °C

Run No.	Composition of load solution	Composition of eluent	*p*H	Observation	$\varepsilon \times 10^5$	Illustration
1	25 ml of 0·1 M UCl_4 + 0·1 M H_3L + 0·1 M H_2SO_4	0·1 M K_3L + 0·1 M K_2SO_4	1·65	Precipitation at rear of band		
2	25 ml of 0·1 M UCl_4 + 0·1 M H_3L + 0·1 M H_2SO_4	0·1 M K_3L + 0·1 M K_2SO_4	1·36	Displacement chromatography	6·0	Fig. 1
3	25 ml of 0·1 M UCl_4 + 0·1 M H_3L + 0·1 M H_2SO_4	0·1 M K_3L + 0·1 M K_2SO_4	0·99	Continuous broadening of band		
4	60 ml of 0·1 M UCl_4 + 0·2 M H_3L + 0·2 M H_2SO_4	0·1 M K_3L + 0·1 M H_3L + 0·1 M K_2SO_4 + 0·1 M H_2SO_4	1·36	Precipitation at rear of band		
5	60 ml of 0·1 M UCl_4 + 0·2 M H_3L + 0·2 M H_2SO_4	0·1 M K_3L + 0·1 M H_3L + 0·1 M K_2SO_4 + 0·1 M H_2SO_4	0·99	Displacement chromatography	5·2	Fig. 2
6	60 ml of 0·1 M UCl_4 + 0·2 M H_3L + 0·2 M H_2SO_4	0·1 M K_3L + 0·1 M H_3L + 0·1 M K_2SO_4 + 0·1 M H_2SO_4	0·65	Continuous broadening of band		
7	60 ml of 0·1 M UCl_4 + 0·5 M H_3L + 0·2 M H_2SO_4	0·1 M K_3L + 0·4 M H_3L + 0·05 M K_2SO_4 + 0·15 M H_2SO_4	0·94	Displacement chromatography	6·2	Fig. 3

M = mol dm^{-3}, L^{3-} = citrate ion

in the sample fraction were oxidized to uranyl ions by addition of concentrated HCl and H_2O_2, and then passed through an anion exchange resin column so that K^+, SO_4^{-2} and citrate ions were removed. Uranyl chloride eluted from the column was ignited to U_3O_8 at around 1050 °C. U_3O_8 was then dissolved in concentrated HCl to prepare a sample for the mass spectrometric analysis.

Results and Discussion

1. *Influence of eluent composition on shape of chromatogram and single stage separation coefficient*

A number of complex formation reactions will take place in uranous solution containing citric and sulfuric acids. Since reliable data on complexation of uranous ions are scarce because of experimental difficulties, it is impossible to accurately estimate the concentrations of the complexes at a given solution composition. However, in view of a few complex formation data so far reported [5], it seems that in the *p*H range 1—2 the complex formation between uranous and citrate ions would be predominant. Therefore, first, we intended to find out the optimum eluent composition for displacement chromatography around the above *p*H range. Seven chromatographic experiments were carried out. The *p*H of the eluents was varied by addition of HCl. The compositions and *p*H values of the elu-

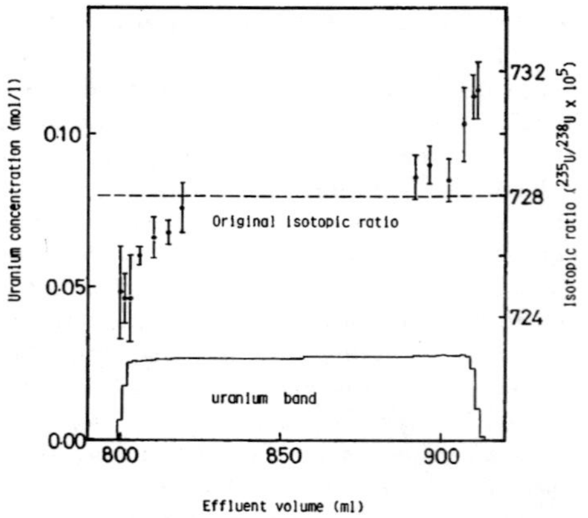

Fig. 2. Elution graph and isotopic distribution for Run No. 5

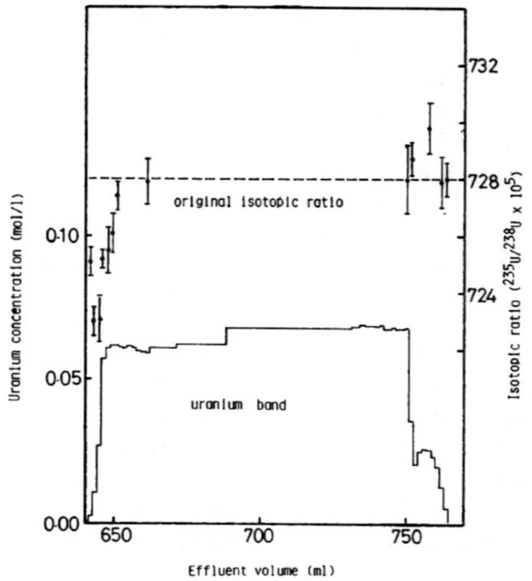

Fig. 1. Elution graph and isotopic distribution for Run No. 2

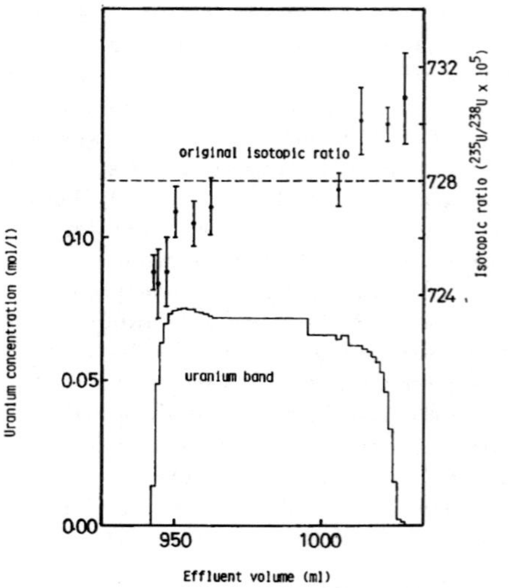

Fig. 3. Elution graph and isotopic distribution for Run No. 7

ents are collected in Tab. 1. The displacement distances of the uranium bands were about 320 cm in all the cases. The temperature of the eluents was kept constant at 50 °C. In runs 1 and 4 uranous ions were precipitated at the rear part of the bands. In runs 3 and 6 continuous band broadening was observed during displacement. In runs 2, 5 and 7 relatively clear displacement chromatograms were obtained. They are illustrated in Fig. 1 to 3, respectively. A small peak at the end of the elution curve in Fig. 1 is the one containing uranyl ions produced by oxidation of uranous ions during displacement. From the above observations it can be said that as far as the compositions given in the present work are concerned, the displacement chromatography is feasible in only a limited *p*H range.

The reactions occuring in the ion exchange column may be described as follows:

(a) At the rear boundary of the uranous band, uranous ions adsorbed in the resin phase are desorbed by complex formation with citrate ions in the external solution, and this desorption reaction of uranous ions are accompanied by adsorption of K^+ and H^+ by the resin phase, which may be expressed as

$$\overline{U^{4+}} + nL^{3-} + mK^+ + (4-m)\,H^+ = \overline{mK^+} + \overline{(4-m)\,H^+} + UL_n^{(4-3n)+} \quad (1)$$

where — denotes the resin phase.

(b) At the frontal boundary of the uranium band, uranous citrate complexes in the external solution are decomposed into uranous and less-citrate-coordinated uranous ions by the presence of hydrogen ions desorbed from the resin, and then uranous and less-coordinated uranous species are adsorbed by the resin, as is expressed by

$$UL_{n'}^{(4-3n')+} + \overline{4\,H^+} = \overline{U^{4+}} + n'L^{3-} + 4\,H^+ \quad (2)$$

When the two reactions (1) and (2) tend to proceed to the right, the uranium band will migrate down to the bottom of the column successfully with sharp boundaries at the both edges of the band, as is shown in Fig. 1 to 3.

(c) Between uranous citrate complexes in the external solution and the uranous ions and less-coordinated uranous species in the resin phase, the isotopic exchange reaction occurs, which is expressed by

$$^{238}UL_n^{(4-3n)+} + \overline{^{235}U^{4+}} = {}^{235}UL_n^{(4-3n)+} + \overline{^{238}U^{4+}} \quad (3)$$

The accumulation of the discrepancy of the equilibrium constant of reaction (3) from unity will result in the observed isotope effects.

The results of the isotopic analyses for the three successful runs are inserted in Fig. 1 to 3. It is clearly seen that ^{238}U is enriched at the front of the bands, while ^{235}U is preferentially accumulated at the rear. This phenomenon is consistent with that observed in the uranous—sulfate system [4], while it is contrary to that observed in the uranyl complexation systems [1]. The reversal has been theoretically predicted by *Kakihana* [6].

The single stage separation coefficients ε for the three runs were calculated as follows. The coefficient ε is defined as

$$\varepsilon = S - 1 = \frac{\overline{[^{235}U]}\,[^{238}U]}{[^{235}U]\,\overline{[^{238}U]}} - 1 \quad (4)$$

where S is the single stage separation factor, [ ] denotes the concentration of isotopes in the aqueous phase, and $\overline{[\;]}$ that in the resin phase. The separation coefficient ε can be calculated with the chromatographic data by means of the following equation [7].

$$\varepsilon = \frac{\Sigma f_i (R_i - R_0)}{Q \times R_0 (1 - R_0)} \quad (5)$$

where f_i is the amount of uranium in the ith fraction, R_0 the original atomic fraction of ^{235}U, R_i the atomic fraction of ^{235}U in the ith fraction and Q the total exchange capacity of the resin bed. The ε values thus calculated were $6{\cdot}0 \times 10^{-5}$, $5{\cdot}2 \times 10^{-5}$ and $6{\cdot}2 \times 10^{-5}$ for runs 2, 5 and 7, respectively. These values are in agreement each other within the experimental accuracy, i.e. the separation coefficient is little affected by the limited change of the composition of the eluent given in the present work. The separation coefficient $\varepsilon = 6 \times 10^{-5}$ at 50 °C obtained in the present work agrees with that ($\varepsilon = 5 \times 10^{-5}$ at 40 °C) found in the uranous — sulfate system [4]. That is to say, the difference in ε between the two systems due to the difference in formation constants are not observed in the uranous systems unlike in the case of the uranyl systems [1]. The separation coefficients estimated in the uranous systems are smaller than those reported in the systems of uranyl complex formation [1] and of electron exchange reactions between U(IV) and U(VI) [8], to say more exactly, they are about two thirds of those in the former systems and less than one tenth of those in the latter.

2. *Influence of temperature on single stage separation coefficient*

So that the influences of the operating temperature on the shape of the chromatogram and the single stage separation coefficient were examined, chromatographic experiments were carried out by changing operating temperature with

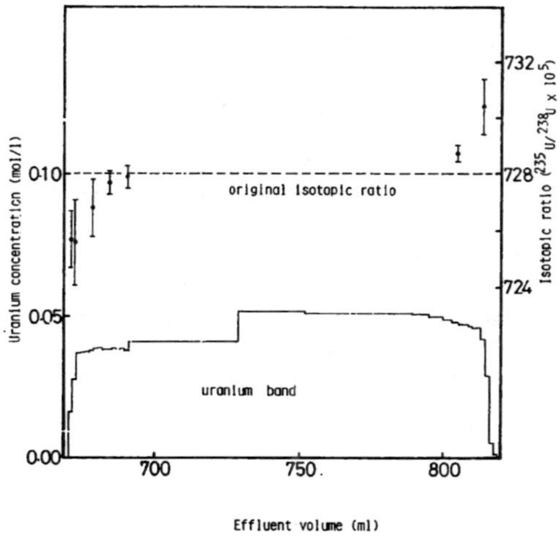

Fig. 4. Elution graph and isotopic distribution for Run No. 8

Tab. 2. Results in chromatography at various temperatures

The compositions of the load and eluent solutions are the same as those for Run No. 5.

Run No.	Temp., °C	$\varepsilon \times 10^5$	Illustration
8	70	4·1	Fig. 4
5	50	5·2	Fig. 2
9	30	6·6	Fig. 5
10	11	10·5	Fig. 6

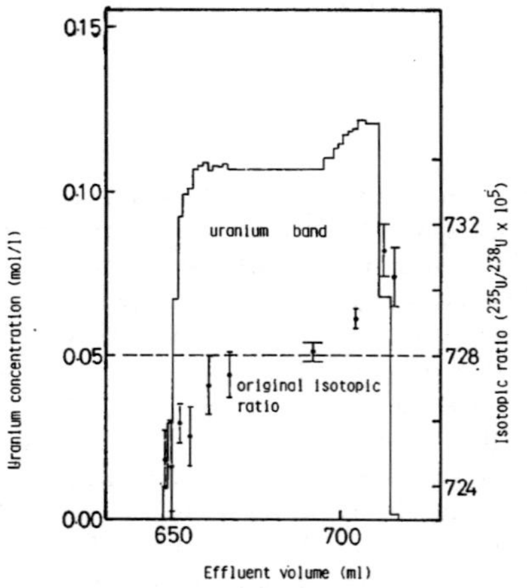

Fig. 5. Elution graph and isotopic distribution for Run No. 9

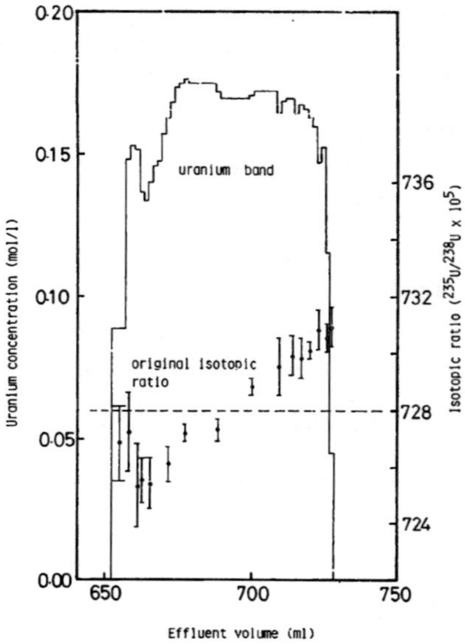

Fig. 6. Elution graph and isotopic distribution for Run No. 10

the composition and *p*H of the eluent being kept constant. The experimental conditions are summarized in Tab. 2. The chromatograms for runs 8 to 10 are shown in Figs. 4 to 6, respectively. The satisfactory displacement chromatograms were obtained in the three runs. The elution curves were effectively influenced by the operating temperature, i.e. the uranium concentrations of the effluent fractions decreased with increasing temperature. This phenomenon may be mainly caused by the change of uranous—citrate complex formation equilibria with temperature. It is apparent that ^{235}U is depleted at the front of the band and enriched at the rear in all the three runs. This observatio ı is in accord with that at 50 °C described in section 1). In conclusion, the direction of uranium isotope enrichment in the elution curve is not affected by operating temperature. The single stage separation coefficients ε were calculated according to the procedures described in section 1). The calculated values are listed in Tab. 2. It is obvious that the ε decreases with the rise of the operating temperature. This trend is compatible with that predicted by the thermodynamical theory of isotope effects [9], although no exact and quantitative explanation is possible because the complex formation equilibria should be influenced by the change of the temperature.

Conclusion

(1) A displacement chromatography in the uranous — citric acid system is feasible in a very limited *p*H range.

(2) Uranium-235 is depleted at the front of the chromatogram and enriched at the rear. This is contrary to the observation in uranyl complexation systems.

(3) The single stage separation coefficient ε is little affected by the change of the compositions of the eluent at a given temperature. The ε value is about 6×10^{-5} at 50 °C.

(4) The single stage separation coefficient ε in the present system is much smaller than those reported in the systems of uranyl complexation and of electron exchange reactions between U(IV) and U(VI).

(5) The single stage separation coefficient ε decreases with rising temperature.

Received January 12, 1983

References

[1] *M. Okamoto* et al., Isotopenpraxis **16** (1980) 293
[2] *Y. Tanaka* et al., J. Inorg. Nucl. Chem. **43** (1981) 3291
[3] *Y. Tanaka* et al., J. Phys. Chem. **86** (1982) 1015
[4] *Y. Sakuma* et al., J. Nucl. Sci. Technol. **18** (1981) 793
[5] (a) *L. G. Sillén, A. E. Martell*, Stability Constants, Spec. Publ. No. 17, 1964; and Suppl. No. 1, Spec. Publ. No. 25, 1971. The Chemical Society, London; (b) *D. D. Perrin*, Stability Constants of Metal-ion Complexes: Part B. Organic Ligands. Pergamon Press 1979
[6] *H. Kakihana*, J. Chim. Phys. **60** (1963) 81
[7] *H. Kakihana, T. Kanzaki*, Bull. Tokyo Inst. Technol. **77** (1969) 90
[8] *Y. Fujii* et al., J. Nucl. Sci. Technol. **15** (1978) 745
[9] *J. Bigeleisen, M. G. Mayer*, J. Chem. Phys. **15** (1947) 261

Isotopenpraxis 19 (1983) 8, S. 265–267

On Labelling with Generator Nuclides*)

R. Otto, P. Hecht

(Academy of Sciences of the GDR, Central Institute of Isotope and Radiation Research)[1])

Short-lived radioisotopes separated from radionuclide generators are widely used. In the paper possibilities of labelling with the daughter nuclides of the commercially available Mo/Tc-, Sn/In- and Te/I-generators and of the self-made Ba/La-generator in industrial tracer experiments are presented. The transfer of the daughter nuclides from the generator eluates into organic phases and the labelling of oil-phases and solid particles are investigated. The developed simple, quick and efficient methods are suitable for routine application under industrial conditions. Some examples of industrial applications of the generator nuclides are given, too.

Keywords

indium 113; industry; iodine 132; lanthanum 140; radioisotope generators; technetium 99; tracer techniques

1. Introduction

For about 20 years industrial tracer experiments have been carried out by the Central Institute of Isotope and Radiation Research. The benefit of such experiments can be very large, because their results allow to realize a higher throughput of the plants; they lead the way to an improvement of the quality of the final products, and the raw material and energy expense can be decreased. In all experiments practicable tracers must be prepared for very different phases and materials. Among other demands it's very important to find a tracer with a suitable half-life and tracers with short half-lives are often desirable. However, the site at which the measurement is required may be remote from an isotope production facility and this circumstance restricts the application to radioisotopes with half-lives greater than some hours. But radionuclide generators represent a continuous source of short-lived products. They can be milked repeatedly. For many years radionuclide generators have been used in medicine and in recent years in the laboratory and in industrial practice, too. Generators are reservoirs of such practicable short-lived isotopes as ^{99m}Tc, ^{113m}In, ^{132}I, and ^{140}La, which permit to carry out consecutive experiments without interference from previous injections, which allow to improve the radiation safety and to eliminate or minimize the waste disposal problem. In what follows possibilities of labelling with the daughter nuclides of the commercially available Mo/Tc-, Sn/In-, and Te/I-generators and of the self-made Ba/La-generator are presented. But generators yield their daughter nuclides in a predetermined chemical form, however, it is most useful to have the daughters in different chemical forms. Therefore, the transfer of the daughter nuclides from the aqueous generator eluates into organic phases and the labelling of oil phases and of solids are described and some examples of industrial applications of generator nuclides are presented.

2. Labelling of organic phases

At first the transfer of the daughter nuclides into organic solvents is described. Mo/Tc- and Te/I-generators form their daughter nuclides as pertechnetate and iodide in an isotonic sodium chloride solution and the ^{113m}In is eluated by 0·05 N hydrochloric acid. And so these generators furnish eluates in a chemical form which requires little or no further preparations before labelling aqueous phases.

In order to label organic phases or solvents the short-lived daughter nuclides must be converted into a form soluble in such phases. The conversions have to be very simple and quick. For this purpose the aqueous eluates can be extracted with organic solvents or solutions. In Tab. 1 some examples are listed. The obtained organic extracts can be directly used to label organic phases. In some cases the extracts must be concentrated before use by evaporating. For example the pertechnetate can be extracted with cyclohexanone or tributyl phosphate. The extraction efficiencies are about 75%, respectively 92%. From alkaline solutions the Tc-activity also can be very effectively removed by methyl ethyl ketone or pyridine.

The eluate of the Sn/In-generator contains the indium as a trivalent cation, which can be extracted with solutions such as 8-hydroxyquinoline/acetyl acetone solved in benzene, di-(2-ethylhexyl)phosphoric acid (HDEHP) in benzene or diethylether saturated with hydrochloric acid.

Evaporating the generator eluate to dryness diethyl ether saturated with hydrochlorid acid for example resolves about one hundred percent of the In-activity.

^{132}I is eluated as an iodide ion. The transfer into organic phases can be easily accomplished by extraction with chloroforme which contains 0·1% of elementary iodine. Or the eluate can be evaporated to dryness, and the dry residue can be dissolved in an ethanolic solution of potassium iodide and elementary iodine. This iodine-132-labelled solution can be used to prepare labelled iodoorganic compounds by isotopic exchange reactions. At the boiling temperature of the solvent and exchange times ranging between 20 and 60 min exchange rates up to more than 90% may be reached. Addition of anhydrous aluminium chloride yields higher exchange rates [1].

This method was tested to provide with such iodine-132-labelled compounds as iodoalkanes and iodocycloalkanes (Tab. 2).

In some cases oil phases must be labelled with nonvolatile tracers. ^{113m}In and ^{140}La were used. The aqueous eluates or extracts were made alkaline and lanthanum oleate or

*) Presented at the 2nd Working Meeting "Radioisotope Application and Radiation Processing in Industry", 28th September to 1st October 1982, Leipzig

[1]) Address: DDR-7050 Leipzig, Permoserstr. 15

Tab. 1. Labelling of Organic Phases

Starting material	Method	Organic phase	Yield
$^{99m}TcO_4^-$ (Saline)	→ Extraction	→ Cyclohexanone	75%
		→ TBP	92%
		→ MEK (from 5 N NaOH)	90%
		→ Pyridine (from 5 N NaOH)	85%
$^{113m}In^{3+}$ (0,05 N HCl)	→ Extraction	→ 8-Hydroxyquinoline/acetyl acetone in benzene	86%
		→ 1% HDEHP in benzene	98%
		→ Diethylether/HCl	90%
	→ Evaporation	→ Diethylether/HCl	~100%
	→ Chem. Reaction	→ Oleate, stearate	~100%
$^{132}I^-$ (Saline)	→ Extraction	→ $CHCl_3 + 0{\cdot}1\%\ I_2$	79%
	→ Evaporation	→ Ethanol	95%
		→ Ethanolic KI/I_2-sol. ↓ Iodoorganic compounds	99%
$^{140}La^{3+}$ (TBP/HNO_3)	→ Extraction	→ Methanol	85–92%
		↓ Oleate, stearate	100%

Tab. 2. Organic Iodine Compounds Labelled with I-132

Iodine isotopic exchange between organic iodine compounds and ^{132}I-labelled KI/I_2 in ethanol, 90%, at b.p.

	F, %	
	20 min	30 min
Iodoethane	93·7	—
1-Iodopropane	91·9	95·6
1-Iodobutane	92·1	96·0
1-Iodohexane	77·9	92·9
1-Iodononane	82·0	—
Iodocyclohexane	72·0	80·9
Diiodomethane	41·9	59·2
Iodobenzene	—	<0·5
1-Iodo-2-methylbutane	63·6	78·5

$$F \text{ (exchange rate)} = \frac{Z_t}{Z_E} \times 100\ [\%]$$

Tab. 3. Labelling of Solid Particles

Solid particles	Nuclide	Method	Yield
Solid particles → Powders, Granules, Chips etc.	→ ^{99m}Tc → (Saline)	Passing H_2S into the acidified eluate	95%
	→ ^{113m}In → (0·05 N HCl)	Stirring in alkaline solutions	47–99%
		↓ Evaporation to dryness	75–97%
	→ ^{132}I → (Saline)	Precipitation as AgI	81–95%
	→ ^{140}La → (Methanolic sol.)	Stirring	82–93%

stearate with a small content of sodium oleate or stearate was added. After evaporation to dryness the residue was directly dissolved in the oil or predissolved in a small volume of toluene or benzene and then this solution was mixed with the oil.

3. Labelling of Solid Particles

Not only liquid but also solid phases and compounds were to be labelled with short-lived isotopes. In Tab. 3 some examples are shown.

Activated coal could be labelled with ^{99m}Tc by passing hydrogen sulphide into the acidified generator eluate which contains the activated coal. Up to 95% of Tc-activity were fixed on the adsorbent.

The labelling of alumina, infusorial earth, coal, glass powder, plastic chips, ashes, sawdust, and other materials with ^{113m}In was accomplished by a simple absorption procedure. The powdery material was added to the alkaline eluates. After some stirring the powder was separated by filtration, or the mixture was evaporated to dryness. In

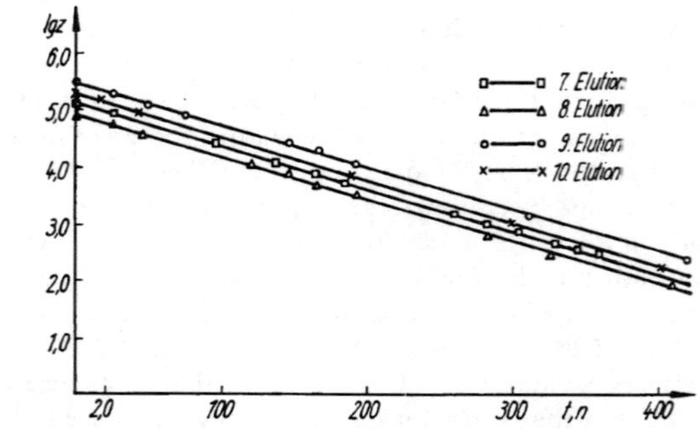

Fig. 1. Decay curves of the ^{140}La-eluates
correction: unstead of *t, n* read *t, h*

most cases 90–95% of the activity were fixed on the material. The labelling of different solid materials with ^{132}I was arranged by precipitating silver iodide in the presence of the solid particles as alumina, infusorial earth, sawdust, straw etc. 95–99% of the activity are fixed.

Solid particles as plastic chips, ceramic materials or dextrin can be labelled with ^{140}La by stirring the methanolic or aqueous solution in the presence of the material. After 10 minutes the powder can be separated by filtration.

In most cases the ^{140}La is separated from a self-made generator. The ^{140}Ba is fixed on a column of alumina. The elution of the ^{140}La is effected by tributyl phosphate saturated with concentrated nitric acid. The generator per-

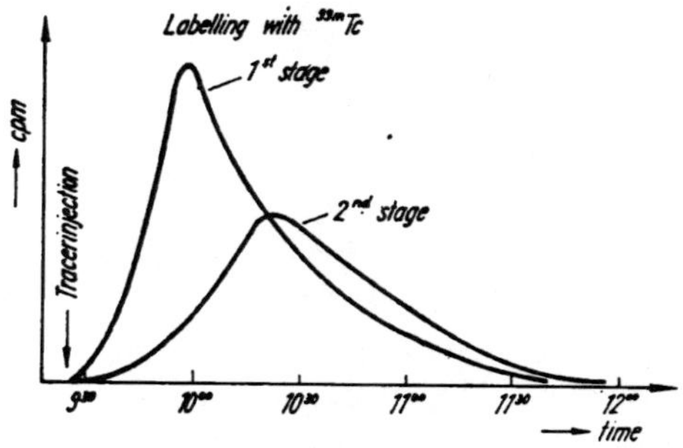

Fig. 2. Thinlayer evaporating equipment — Residence time density function of the bottoms of the 1st and 2nd distilling stage

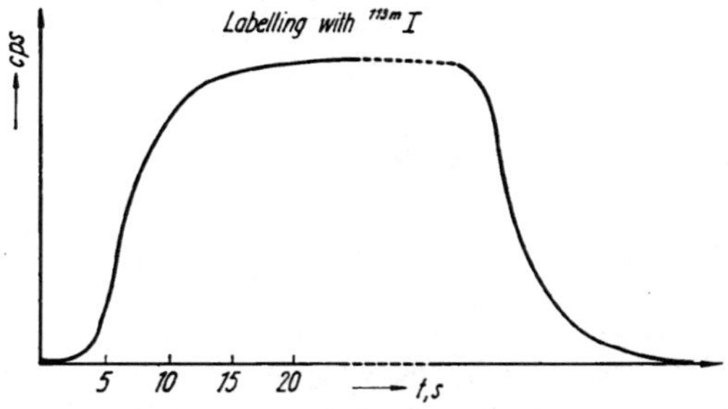

Fig. 3. Chemical reactor — Residence time distribution function of the oil phase
correction: instead of ^{113m}I read ^{113m}In

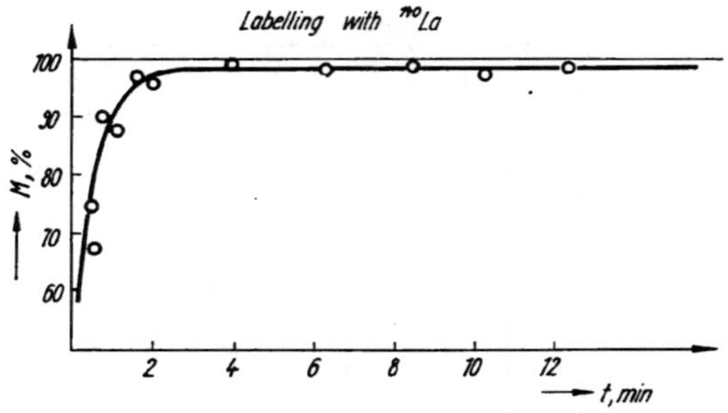

Fig. 4. Batch mixer — Mixing efficiency as a function of mixing time Sabelled binder
M (Mixing efficiency) = (1-*V*) 100, %; *V* = coefficient of variation

forms at 85% efficiency. In our case the contamination of ^{140}La by ^{140}Ba in the eluate is less than 0·01%. After 10 elutions the purity of the eluate didn't decrease (Fig. 1). If the necessary daughter nuclide can be reextracted by a small volume of water, only 1—5% of the activity remains in the tributyl phosphate. Furthermore the water can be evaporated to dryness and the activity can be solved in methanol and now the methanolic solution is suitable to label some organic phases with a neutral tracer.

4. Generator Applications

In recent years bringing to perfection the labelling ,measuring and computer techniques the use of the very practicable generator nuclides has more and more increased. The short-lived generator nuclides have very useful industrial applications such as residence time and flow rate measurements, mixing studies, determinations of material balances and leak detections. In what follows some examples are shown:

1) In order to improve the energy balance and to improve the quality of the final product it was required to measure the residence time of the nonvolatile battoms in the 1st and 2nd stage of a large-scale still (Fig. 2). Before pumping the raw product into the first stage an aqueous ^{99m}Tc-generator eluate was injected. Under operating conditions samples of the bottoms were collected and their specific counting rates measured. The results led the way to improved operating conditions, and now a final product of constant and high quality is put out, even if the quality of the raw product decreases.

2) Another example: At a pilot plant operated at high temperature and pressure the residence time distribution of the oil phase was to be measured, because the optimal operating conditions were to be found out. In/La-oleate was prepared as described, dissolved in the oil and the labelled oil was pumped into the reactor. The obtained curve is shown in Fig. 3. The results accomplished the know-how to design a large-scale plant.

3) And now a third example: In a plant producing abrasive tools the homogeneity of the mixed solid components was to be investigated. The ceramic binder was labelled by stirring with a methanolic solution of ^{140}La. By sampling and measuring the specific counting rates of the samples the mixing efficiency as a function of the mixing time was obtained (Fig. 4). In accordance with the results the mixing conditions could be significantly improved.

At the end of their paper the authors beg pardon for making another attempt to demonstrate the usefulness of short-lived isotopes separated from radionuclide generators.

Received January 5, 1983

References

[1] *R. Otto*, Isotopenpraxis **12** (1976) 22

Isotopenpraxis 19 (1983) 8, S. 268–269

Apparatus for Gamma Activation Analysis*)

I. Pavlicsek, V. Stenger, A. Veres

(Institute of Isotopes, Budapest)[1])

An irradiation device containing 740 TBq (20,000 Ci) ^{60}Co giving 7·74 × 10^3 C/kg intensity (30 MR/h) and a pneumatic dispatch rabbit system controlled by an automatic system provided with an appropriate programme was built. The Se, Br, Sr, Ag, Cd, In, Hf, Ir, Au and Pt content of samples of different origin can be determined in a volume of 5 cm^3 with 0.05–1 mg/g sensitivity.

Keywords

activation analysis; activation energy; gamma sources; isomeric nuclei; metastable states

1. Introduction

It is known that irradiating a few elements with γ-rays of energy above 1 MeV one of their isotopes having isomer state becomes radioactive. The radioactivity is proportional to the quantity of the element irradiated and this fact can be used for non-destructive activation analysis of elements Se, Br, Sr, Ag, Cd, In, Hf, Ir, Au and Pt. In Tab. 1 we give a few essential characteristics of the isotopes of these elements.

We will not deal with the theoretical problems of [γ, γ′]-reactions, neither will study the inelastic scattering of γ-rays on these nuclei, because many studies have been made to determine the activation cross sections of the isomeric states using accelerator bremsstrahlung or monoenergetic γ-rays from radioactive decay. E.g. photoactivation of the isomeric states was carried out by *Pontecorvo* and *Lazard* [1] by means of bremsstrahlung; *Harbottle* had got measurable isomeric activity for ^{115}In and ^{111}Cd using 66·6 TBq (1800 Ci) ^{60}Co and 48·1 TBq (1300 Ci) ^{182}Ta sources [2]. Later, many investigations were made to determine nuclear parameters and many publications appeared on the practical use of isomer activation.

It was expected that further improvement of the method would be expected from improving selectivity and, first of all, rapidity of detection. Rapidity of detection is limited by the very short half life (5–44 s) of isomeric states of most nuclei mentioned above (7 from among 12). The short half-life demands transport of the investigated material from irradiation point to detector system as rapid, as possible. The problem is augmented by the fact that detectors must be saved from the radiation of the source because of the very low level activation to be detected (e.g. the integrated cross section using ^{60}Co sources amounts to about 10^{-33}–10^{-31} cm^2).

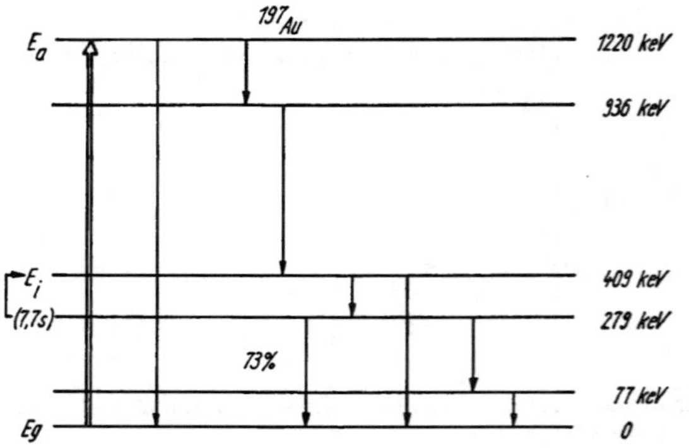

Fig. 1. Energy diagram showing the excitation of the isomeric states via the first activation level of ^{197}Au
E_a, E_i, E_g are the activation level, isomeric state and ground level, respectively

The mechanism of isomer activation can well be understood from Fig. 1. The Compton scattered γ-rays of the source excite the nucleus to the activation level E_a from which the isomeric state is reached directly or by cascade transition [3–5]. The first activation level of ^{197}Au is

*) Presented at the 2nd Working Meeting "Radioisotope Application and Radiation Processing in Industry", 28th September to 1st October 1982, Leipzig

[1]) Address: 1525 Budapest, P.O.B. 77, Hungary

Tab. 1. Data of nuclei having isomeric states

nucleus	^{77}Se	^{79}Br	^{87}Sr	^{107}Ag	^{109}Ag	^{111}Cd	^{115}In	^{179}Hf	^{191}Ir	^{197}Au	^{195}Pt
abundance, %	7·58	50·7	7·0	51·8	48·2	12·8	95·7	13·7	37·3	100	33·8
radiation γ-energy, keV	161·9	207·2	388·3	93·1	88·0	149·6 246	336	161 217	129	278·5 130·5	130 99 31
half-life	17·4 s	4·86 s	2·8 h	44 s	40 s	48·6 min	4·5 h	19 s	49 s	7·7 s	4·0 d

Tab. 2. Sensitivity of the apparatus

elements	Se	Br	Sr	Ag	In	Ir	Au	Pt	Cd	Hf
mg/5 cm³	0.08	0.03	0.8	0.4	0.03	0.09	0.7	0.8	0.4	0.06

1220 keV and the isomeric activation level E_i is 409 keV. This level, having 7.7 s half-life, makes the measurement of the γ-activity of ^{197m}Au possible.

2. Experimentals

The irradiation and the measuring arrangement can be seen in Fig. 2. A water pool was used for the storage of the 740 TBq (20 kCi) ^{60}Co sources. The pool has many advantages: e.g. the 5 m water shielding results in a very low background at the measuring point, placed not far from the pool (about 2 m), and we can easily join the sources with a pneumatic rabbit system.

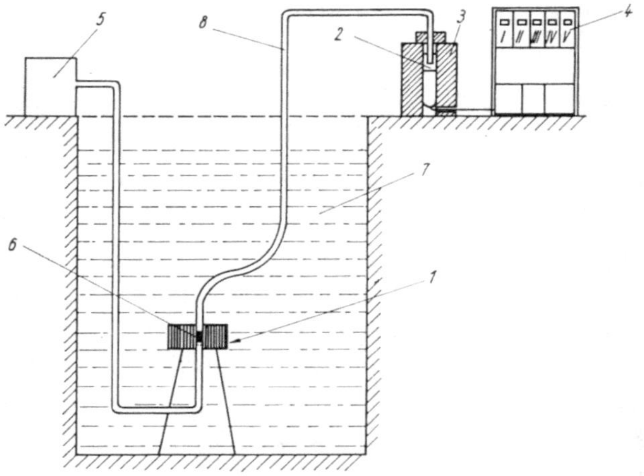

Fig. 2. Irradiation and measuring apparatus

1 – ^{60}Co sources; 2 – detector; 3 – shielding; 4 – electronic and control device; 5 – pneumatic system; 6 – sample; 7 – water shielding; 8 – tube of the dispatch rabbit system

The next principal part of the gamma-activation rabbit system is the detector. For measuring the very low activity of the samples a well-type NaI(Tl) detector system and crystal mounted multiplier were used, with 5 cm lead and 10 cm iron shielding and a 2 mm copper inner coating. Thus we achieved a very low background in the 40–400 keV energy interval where the γ-energies of the isomeric activity were expected. The integral count of that region amounted to 260 cpm.

The period of a single analysis using a pneumatic rabbit system with 1–2 s running time depended on the half-life of the isomeric state. The running time was measured by two photodiode indicators, placed at the detector and near the sources, respectively. The sources and the detector were joined with a 8 m long brass tube, and a pneumatic system yielded pression and depression, respectively, to move the sample holder between them to and fro. The 5 cm³ sample holder was made of aluminium. The speed of transport depended on the density of the substance to be analysed. The analyser was suitable to measure the activity of the samples after irradiation by counting the background, controlling the irradiation time and measurement cycle, and evaluating data by means of a simple programme which made corrections for irradiation time, decay time and measurement period (in case of manyfold cycling). When there is very small amount of substance to be measured or a very low concentration, one cycle is not enough to yield suitable data, so the irradiation – measurement cycle can be preset to 20–50 cyclings and carried out automatically.

3. Results

Having one hour for measuring with various cycle numbers and irradiation times sensitivity changed according to Table 2.

The main advantage of the gamma-activation method is its simplicity and rapidity and that it gives information on the composition from deeper layers than other methods do, such as XRF or AAS (Atomic Absorption Spectrometry) which are suitable for surface analysis.

We have analysed ancient coins for numismatists to determine their Au and Ag content and Au/Ag ratio and applied the method to the rapid determination of Au content in minerals, too.

Received January 10, 1983

References

[1] *B. Pontecorvo, A. Lazard*, Compt. Rend. 208 (1939) 99
[2] *G. Harbottle*, Nucleonics 12 (1954) No. 4, p. 64
[3] *A. Veres, I. Pavlicsek*, Trace Element Analytical Chemistry in Medicine and Biology. Berlin – New York: Walter de Gruyter 1980, p. 393
[4] *A. Veres*, At. Energ. Rev. 18 (1980) 217
[5] *A. Veres, I. Pavlicsek*, Journal of Rad. Chem. 69 (1982) 325

Isotopenpraxis **19** (1983) 8, S. 270–272

Determination of Calcium and Iron and Measurements of Ash Content in the Brown Coal*)

P. Urbański, M. Jankowska, E. Kowalska

(Institute of Nuclear Research, Warszawa)[1])

D. Wagner

(Central Institute of Isotope and Radiation Research, Academy of Sciences of the GDR, Leipzig)[2])

Applying the XRF technique and the low-energy X-ray backscattering method an attempt of calcium and iron determination, as well as measurement of the ash content in the brown coal was made. A ^{238}Pu source, an argon filled proportional counter and a three channel pulse height analyser were used. A simple theoretical model is proposed and obtained results are discussed.

Keywords

ash; brown coal; calcium; iron; proportional counter; XRF

1. Introduction

One of the most widely used method of ash content determination in coal is low energy gamma or X-rays backscattering technique with simultaneous measurement of X-ray fluorescence excited in iron. An aluminium filter of appropriate thickness plcaed on the detector window allows proper ratio of the backscattered and fluorescence intensity to be registered by the detector. This method applied for several years for ash determination in the black coal [2–5] exhibits however a few disadvantages, from which the most important are:

- Compensation of the influence of iron content variations on the results of ash content measurements is valid only in limited range of iron content;
- The method does not assure compensation of the calcium content influence on results of ash determination. This matter may be not so important for black coal as for the brown one.

In presented paper an attempt is made to propose such a technique, which could allow to perform ash content measurements in wider range of iron content, as well as for varying calcium content in the brown coal. A simple theoretical model is proposed and obtained results are discussed.

2. Theoretical model

Fig. 1 shows the spectrum of secondary radiation scattered and excited in powdered sample of the brown coal. The sample was irradiated by ^{238}Pu X-rays, and the secondary radiation was detected by an argon filled proportional counter. Three characteristic areas can be distinguished in the presented spectrum: X-ray fluorescence radiation excited in calcium I_{Ca} and in iron I_{Fe}, as well as scattered radiation I_r. The basic information about ash content is carried by the intensity of scattered radiation. However, I_r also acutely depends on the iron and calcium concentration in the sample.

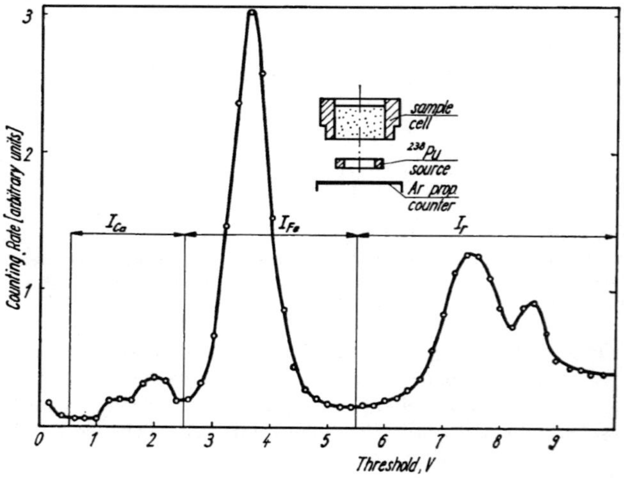

Fig. 1. Spectrum of Secondary radiation from the brown coal sample

The main aim to be achieved is to find such a function

$$f(I) = f(I_r, I_{Ca}, I_{Fe}) \tag{1}$$

depended on the ash content in the coal, which also could fulfill the following condition:

$$\frac{\partial f(I)}{\partial w_{Fe}} = \frac{\partial f(I)}{\partial w_{Ca}} = \frac{\partial f(I)}{\partial w_{Si}} = \dots = \frac{\partial f(I)}{\partial w_i} \tag{2}$$

where: w_i – weight concentration of an element i in the brown coal sample.

For the practical purposes it may be assumed, that ash consists of the basic component – SiO_2 and the interfering components CaO and Fe_2O_3. So, the expression (2) may be simplified:

$$\frac{\partial f(I)}{\partial w_{Fe_2O_3}} = \frac{\partial f(I)}{\partial w_{CaO}} = \frac{\partial f(I)}{\partial w_{SiO_2}} \tag{3}$$

*) Presented at the 2nd Working Meeting "Radioisotope Application and Radiation Processing in Industry", 28th September to 1st October 1982, Leipzig

Addresses: [1]) 03-195 Warszawa, ul. Dorodna 16, Poland; [2]) DDR-7050 Leipzig, Permoserstr. 15

Intensity of the radiation scattered from a multielement sample can be expressed as (2):

$$I_r = kI_0 \frac{\sum_i (\sigma_{con} + \sigma_c)\, w_i}{\sum_i \mu_i w_i} \qquad (4)$$

where: k — a constant
I_0 — intensity of primary radiation on the sample surface
σ_{con}, σ_c — mass coefficients of coherent and Compton scattering respectively
μ_i — mass attenuation coefficient of an element i in the sample
w_i — weight concentration of an element i in the sample.

The expression (4) is valid for relatively low energy of the primary radiation, when difference between energies of coherent and *Compton* scattered radiation is negligible.

Taking into account, that range and absolute values of the mass scattering coefficients for the most elements in the brown coal are much smaller than those of the mass attenuation coefficients, and assuming that the ash consists only of SiO_2, CaO and Fe_2O_3, the formula (4) may be expressed:

$$\frac{1}{I_r} \sim \alpha_{SiO_2} w_{SiO_2} + \alpha_{CaO} w_{CaO} + \alpha_{Fe_2O_3} w_{Fe_2O_3} \qquad (5)$$

where: α_{SiO_2}, α_{CaO}, $\alpha_{Fe_2O_3}$ — correction coefficients
w_{SiO_2}, w_{CaO}, $w_{Fe_2O_3}$ — weight concentration of the apropriate components in the coal sample

It is easy to find, that function f(I) expressed in the following way:

$$f(I) = \frac{1}{I_r} - (\alpha_{Fe_2O_3} - \alpha_{SiO_2})\, w_{Fe_2O_3} - (\alpha_{CaO} - \alpha_{SiO_2})\, w_{CaO} \qquad (6)$$

meets the condition [3].

As it follows from expression (6) prior f(I) determination, the weight concentration of Fe_2O_3 and CaO in the sample as well as correction coefficients α should be found. The iron and calcium content can be found from I_{Fe} and I_{Ca} measurements, whereas correction coefficient values are constant and may be determined by measurements of specially prepared sets of standard samples.

3. Measurements conditions

All measurements were performed using ^{238}Pu annular source of activity 1·11 GBq (30 mCi) and argon filled proportional counter. Three channel analyser was used, so the intensities I_{Ca}, I_{Fe} and I_r (Fig. 1) were simultaneously registered. The air dry coal sample was powdered so that maximum grain size did not exceed 0·1 mm. The sample cell with mylar foil bottom had the volume of about 40 ml.

Two sets of 42 standard samples were prepared in such a way, that for the basic coal sample CaO and Fe_2O_3 (the first set) or CaO and SiO_2 (the second set) were added. In the first set CaO content varied from 1 to 7%, Fe_2O_3 from 1·4 to 10%, and ash content A^d from 6 to 20%. In the second set, CaO content varied from 1 to 8%, SiO_2 from 1·2 to 20% and A^d from 6·8 to 32%.

4. Results

Determination of CaO and Fe_2O_3

Determination of iron in the brown coal sample by XRF technique does not create severe difficulties. The Fe K peak is well distinguished in the spectrum (Fig. 1), and the only problem to be solved is the matrix effect. However, calcium determination is much more difficult task. The Ca K peak interferes with Fe K escape peak in argon proportional counter, and the matrix effect in case of calcium is much stronger than in the case of iron.

Applying regression analysis procedure for results of standard samples measurements, the following formulas for Fe_2O_3 and CaO weight content in the brown coal were found:

$$w_{Fe_2O_3} = \frac{A_1 I_{Fe} + B_1 I_{Fe} \frac{I_{Ca}}{I_r} + C_1}{1 + D I_{Fe}} \qquad (7)$$

$$w_{CaO} = \frac{A_2 \frac{I_{Ca}}{I_r} + B_2 I_{Fe} + C_2}{1 + D I_{Fe}} \qquad (8)$$

The standard deviation of calcium determination was 0·25% CaO and iron below 0·1% Fe_2O_3.

Determination of ash content

As it was mentioned earlier, the first task in the proposed ash content measurement procedure is determination of the

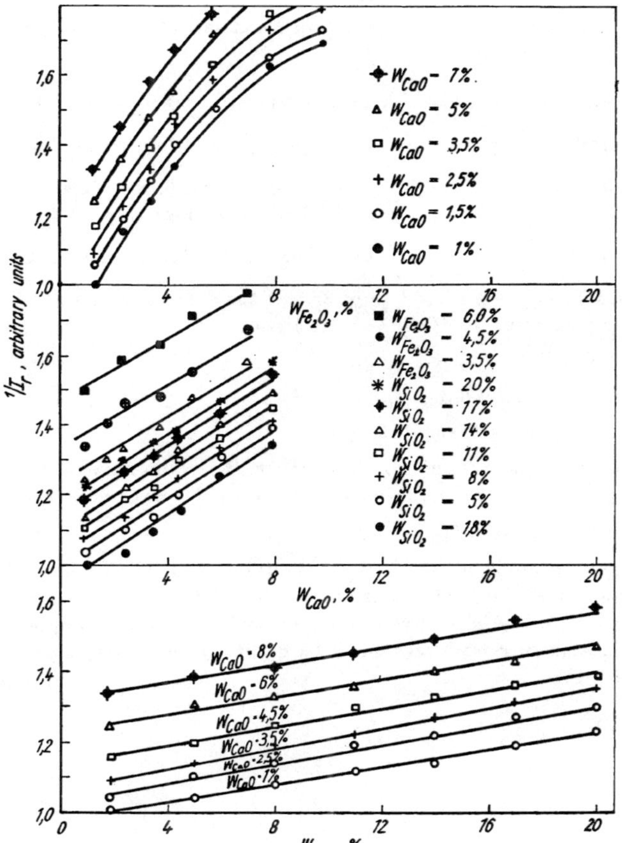

Fig. 2. Influence of SiO_2, CaO and Fe_2O_3 content on intensity of scattered radiation

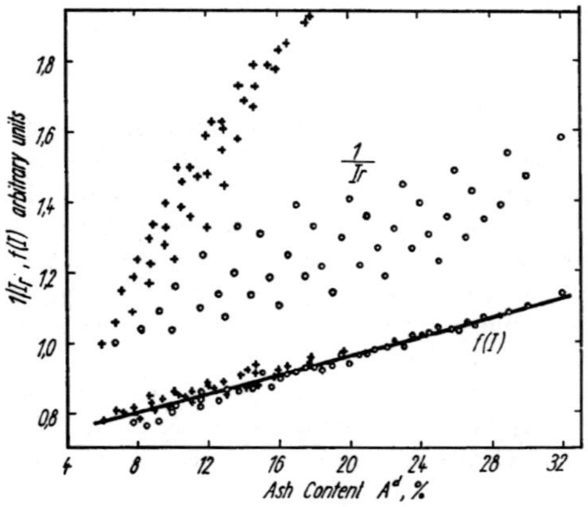

Fig. 3. Influence of the Applied correction on the intensity of scattered radiation

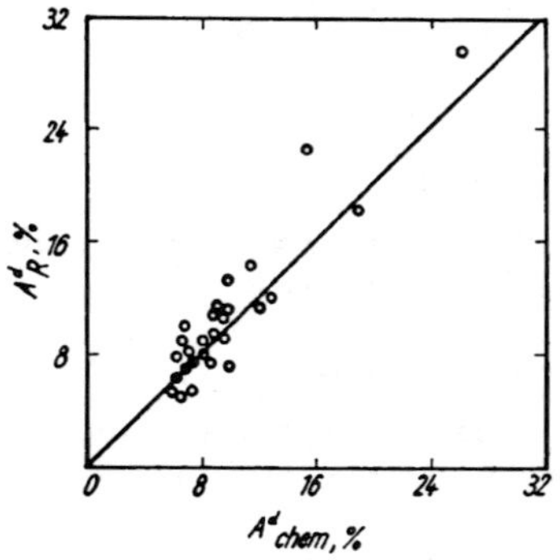

Fig. 4. Comparison of radiometric (A^d_R) and chemical (A^d_{chem}) results of asm determination

correction coefficients values. The value of α_{SiO_2} may be found as slope of the $\frac{1}{I_r} = f(w_{SiO_2})$ function, providing that w_{CaO} and $w_{Fe_2O_3}$ remain constant. The values of α_{CaO} and α_{SiO_2} can be calculated in the similar way.

The characteristics $\frac{1}{I_r}$ versus concentration of SiO_2, CaO and Fe_2O_3 were ploted (Fig. 2) and the following values of the correction coefficients were found:

$$\alpha_{SiO_2} = 0{\cdot}013 \pm 0{\cdot}001$$
$$\alpha_{CaO} = 0{\cdot}051 \pm 0{\cdot}002$$
$$\alpha_{Fe_2O_3} = 0{\cdot}120 \pm 0{\cdot}004$$

Taking into account nonlinearity of the characteristic for iron (Fig. 2) the $\alpha_{Fe_2O_3}$ coefficient may be expressed as:

$$\alpha_{Fe_2O_3} = 0{\cdot}150 - 0{\cdot}006\ W_{Fe_2O_3}$$

Using calculated values of the coefficients α, the intensity of scattered radiation was corrected and results are shown on Fig. 3. The crosses represent results obtained for samples with varying CaO and Fe_2O_3 content, while circles those with SiO_2 and CaO.

Fig. 4 shows results of ash content determination in the samples of the brown coal taken from one particular mine. Standard deviation of the Fe_2O_3 and CaO determination in these samples using algorithmus (7) and (8) was 0·12% and 0·20%.

5. Discussion

The results of standard samples measurements generally confirm supositions made in the theoretical consideration. The only exception is nonlinearity of the characteristic $\frac{1}{I_r}$ versus Fe_2O_3 content for higher iron concentration. However, this characteristics is linear up to about 5% Fe_2O_3 and higher iron contents in the brown coal are rather seldom met.

The values of α coefficients show influence of varying content of the calcium and iron oxides on the intensity of scattered radiation. If this intensity is used for ash content measurements, and assuming that ash consists of silica only, one can easily calculate, that change in CaO content in coal of 1% results in 4% change of the ash content, whereas 1% of Fe_2O_3 change correspond to 9% of the ash content. The above figures show the necessity of I_r correction not only for iron but also for calcium content in the brown coal.

The proposed correction technique allows significant reduction in influence of the iron and calcium content on the ash determination in the brown coal. Moreover, the additional advantage of the presented technique is linearity of the analytical function $f(I)$ versus ash content in the coal.

6. Conclusion

A fast, nondestructive method of ash, as well as calcium and iron content determination in the brown coal may be realised using relatively simple, three channel analyser.

Due to the significant influence of the CaO content on the intensity of scattered radiation, to obtain reliable results, a correction of scattered intensity not only for iron, but also for calcium content is necessary. Since energy of Ca K radiation is low, and its measurements in severe industrial conditions can cause some difficulties, the proposed method is generally intended for laboratory purposes.

The correction procedure needs performing some simple calculations, but it is planned to build measuring equipment based on microprocessor technique which will provide also all necessary computation facilities.

Received January 6, 1983

References

[1] *L. T. Kiss*, Analytical Chemistry 38 (1966) 1731
[2] *J. R. Rhodes, J. C. Daglish, C. G. Clayton*, in: Radioisotope Instruments in Industry and Geophysics 1 (IAEA, Vienna) (1966) 447
[3] *I. S. Boyce, C. G. Clayton, D. Page*, in Nuclear Techniques and Mineral Resources, IAEA, Vienna 1977, 135
[4] *C. G. Clayton*, ibid. 85
[5] *C. G. Clayton, M. R. Wormald*, International conference on Industrial Application of Radioisotopes and Radiation Technology, Grenoble (1981) IAEA-CN-40/119

Kurze Originalmitteilungen

Isotopenpraxis 19 (1983) 8, S. 273–274

The Influence of Airing Systems on the Residence Time in Biological Clarification Plants*)

H.-G. Könnecke, D. Luther

(Central Institute of Isotope and Radiation Research, Academy of Sciences of the GDR)[1])

Keywords

dynamic function studies; flow rates; technical specifications; tracer techniques; ventilation; waste processing; water pollution

The aim of the study was to examine the influence of different airing systems in biological cleaning plants on the residence time distributions of the aqueous phases using radiotracer techniques. The scheme of the plants are given in Fig. 1. In the first plant airing is realized by an impeller. The characteristics of this plant are: volume of the pool $V_1 = 7700\,m^3$; throughput of sewage $D_1 = 18\,000\,m^3/d$. In the second plant (eight pools; volume of one pool $V_2 = 4000\,m^3$; throughput of the whole plant $D_2 = 100\,000\,m^3/d$) airing tubes, installed on the lengthside of the pool, are used.

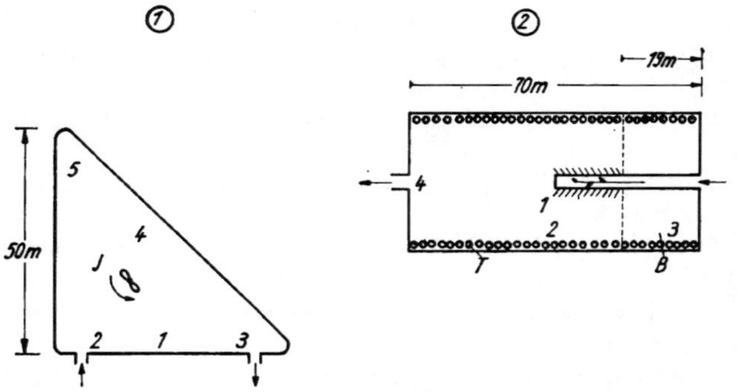

Fig. 1. Scheme of both the inquired biological clarification pools
1 ... 5 = measuring positions; T = airing tubes; I = airing impeller; B = biological activation of the sludge phase

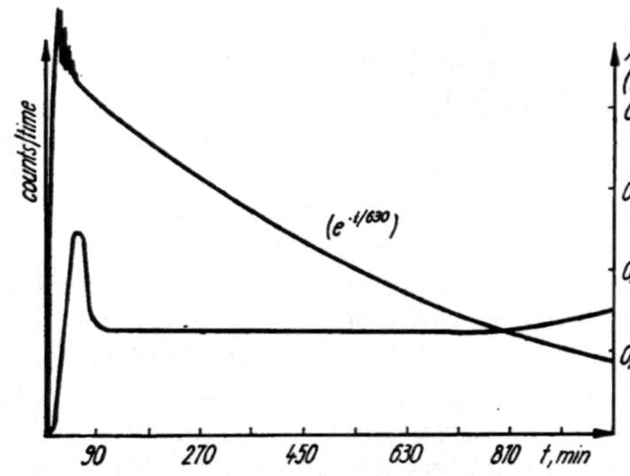

Fig. 2. Residence time spectrum and intensity function of pool 1

In order to determine the hydrodynamic characteristics of the pools the intensity functions are calculated by the equation (2)

$$\lambda(t) = \frac{E(t)}{1 - F(t)}$$

The residence time spectrum and the intensity function of the first examined pool, equipped with an airing impeller, are given in Fig. 2 and Fig. 3. The mean residence time of the aqueous phase in this pool was calculated to be 630 min. The identified by-pass flow (Fig. 3) was reduced by returning the impeller's direction of rotation.

The residence time spectrum of the second pool, equipped with airing tubes (Fig. 4), is type ideal mixing reactor (3).

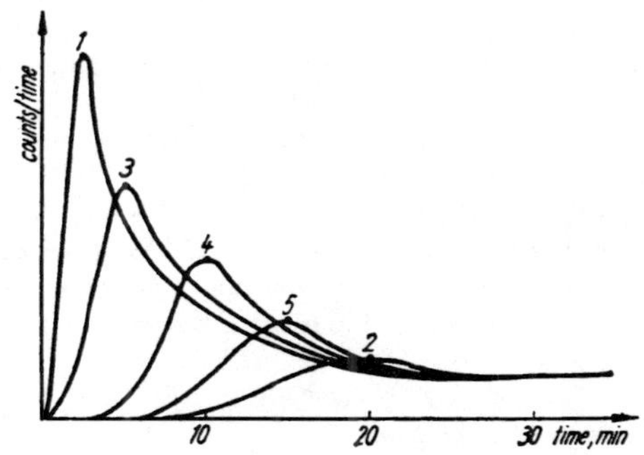

Fig. 3. Residence time spectra of pool 1 (counters 1 to 5)

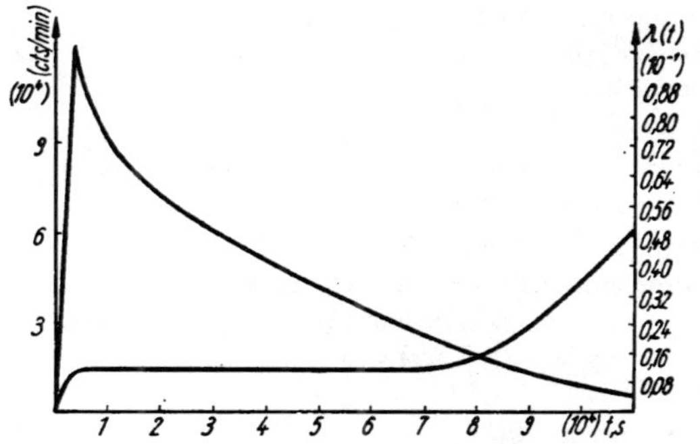

Fig. 4. Residence time spectrum and intensity function of the type ideal mixing reactor, pool 2

*) Presented at the 2nd Working Meeting "Radioisotope Application and Radiation Processing in Industry", 28th September to 1st October 1982, Leipzig

[1]) Address: DDR-7050 Leipzig, Permoserstr. 15

The mean residence time in this plant was found to be abnormal with respect to the throughput. If the throughput in one pool is 625 m^3/h, the mean residence time is 374 min while the mean residence time is only 355 min when the throughput was reduced to 125 m^3/h.

Regarding this abnormal residence time behaviour the sewage input was displaced to the frontside of the pool. This action leads to an increasing throughput of 20%, i.e. the reduction of the necessary energy for airing of 10^6 kWh/a for the whole plant.

It is shown that clarification plants can be optimized by examing the hydrodynamic characteristics of the aqueous phases in pools with different airing systems.

Received January 14, 1983

References

[1] *J. Schön*, 2nd Working Meeting "Radioisotope Application and Radiation Processing in Industry", Leipzig, 28th September to 1st October 1982, Paper No. 5

[2] *P. Noar, R. Shinnar*, IEC Fundamentals 1963, 278

[3] *J. S. Charlton* et al., Radioisotope Techniques for the Investigation of Problems in the Chemical Industry, Grenoble, 28th September to 2nd October 1981, IAEA-CN-40/110

Isotopenpraxis 19 (1983) 8, S. 274–276

Solid Phase Peptide Synthesis of ^{125}I Labelled Insulin Fragments
Part. 1: Synthesis of ^{125}I Labelled Insulin Fragment B_{3-6}-Asparagine—Glutamine—Histidine—Leucine—OCH_3

Y. M. Megahed, M. F. Abdel-Wahab, M. A. Zewall, A. M. Shalaby

(Radioisotope Department Atomic Energy Establishment,
and Tanningmaterial and Protein Laboratory National Research Centre, Cairo, Egypt)

Keywords

asparagine; glutamine; histidine; insuline; iodine 125; labelling; leucine; peptides

Introduction

In recent years, the preparation of several biologically active peptides did impressive successes [1,2]. With the development of new reagents and techniques, the synthesis of some small peptides became an easy reach [3]. However these procedures are not idealy fit to the synthesis of long chain Polypeptide chains, because of the technical difficulties, concerning solubility and purification, as the number of amino acid residue increases.

A new approach to radiopeptide synthesis is now done in an effort to overcome such difficulties.

The present work deals with the basic idea of the solid phase method [4] and with labelling of the prepared tetrapeptide by nascent ^{125}I to facilitate the identification and purification of the tetrapeptide insulin fragment (B_{3-6}).

Materials and Methods

- Starting resin polystyrine — 2% divinyl benzene copolymer beads. Bio-Rad (Bio-beads S-X_2); (Merrifeld 1964) [5], from Fluka
- Protecting group used as (i) Boc-Leu-O-P)
  Boc-His-(Ztf)-DHA dichlorohexylammonium salt Boc-Glu (Mbh)
- Boc-Asp (Mbh)
- Chromatography paper, with the n-butanol—acetic acid—water (4:1:5) system
- N HCl/AcOH for deprotection
- Iodinating mixture of 1 mCi $Na^{125}I$ (Amersham) in 0·02 ml phosphate buffer *p*H 7·5, 50 µg Chloramine-T
- Sephadex G-50 gel-filtration column

Experimental

Boc-Asp (Mbh), Glu (Mbh) Hist (Ztf) Leu-OCH_3 has been used according the following steps:

Attachment of Boc-Leu to resin as described by Merrifield [5] using 1·52 g of Boc-Leu-OH (6·56 mmol).

Cleavage of Boc-group from Boc-Leu resin, using 15 ml of 1 N HCl for 30 min.

Coupling of Boc-His (Ztf) to leu resin using Boc-His. (Zft) DCHA in acid medium.

Cleavage of Boc group from Boc-Hist (Zft) Leu-O-P., using 1 N HCl/AcOH.

Coupling of Boc-Glutamine (Mbh) to the dipeptide His-(Ztf).

Leu-O-P, using N-hyroxy benzotriazole to avoid any possible side reaction.

Cleavage of (Boc-t-butyl-oxy-carbony) group from Boc-Glu-(Mbh) His (Ztf) Leu-O-P using 1 N HCl/AcOH and 10% Triethylamine in methylchloride (Et_3N/CH_2Cl_2):

Coupling of Boc-Asparagin (Mbh) to the tripeptide Glu (Mbh) Hist (Ztf) Leu-O-P-, using 0·87 g Boc-Asp (Mbh) in methylene chloride.

Cleavage of tetrapeptide-Boc-Asp-(Mbh) Glu (Mbh) His (Ztf)-Leu-O-P from polymer applying *Stewart* method [6] using 20% triethyl amine/methyl alcohol. The tetrapeptide was recrystallised from ligroin (40–60°) in a yield of (87·8%) with respect to the first amino acid attached to resin, melting point (198–202 °C). B-Radioiodination of the tetrapeptide: by ^{125}I was carried out applying Chloramin-T-method with some modification using 1 mCi ^{125}I, in phosphate buffer (*p*H 7·4) anf 50 µg Chloramin-T as iodinating mixture. The tetrapeptide (22·5 mg) was then added as solvent to the iodinating mixture, let to react for 1 min add. metabisulfite (120 µg) and lcc followed KI. Purification of the labelled tetrapeptide was carried out applying our method [8] then identification was done applying radiochromatography, localization by (nin-hydrine, benzedine developer and radioscanning by using automatic radiochromatogram scanner. To confirm the results amino acid analysis (Fig. 4) was achieved.

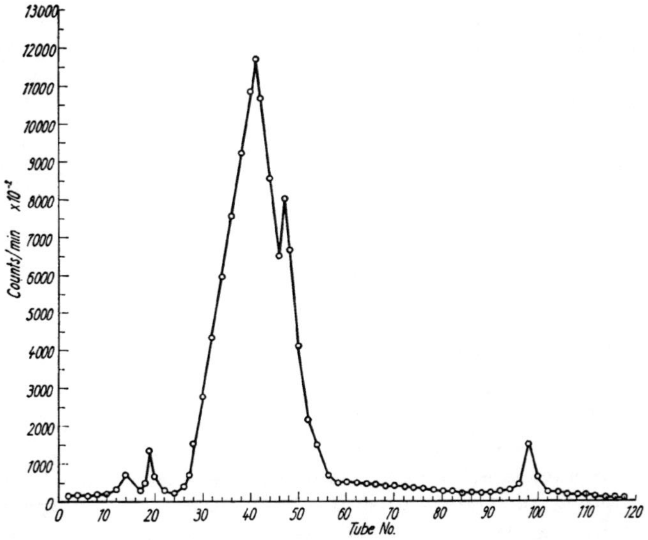

Fig. 1. Fractionation of ^{125}I labelled peptide fragment [insulin (B_{3-6})] on sephadex G-50 column
Radioactivity assay of eluted fractions

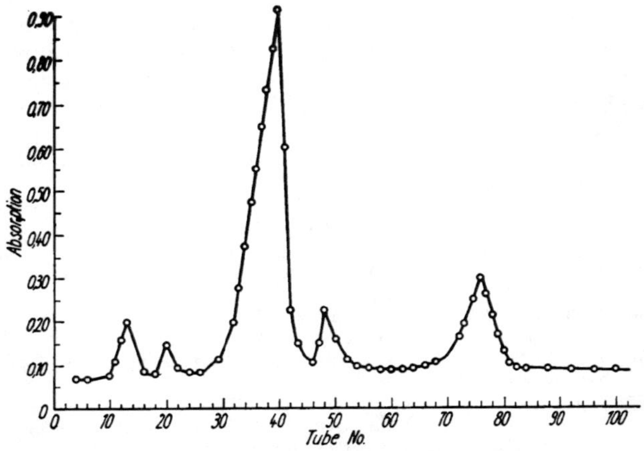

Fig. 2. Fractionation of ^{125}I labelled peptide fragment [insulin (B_{3-6})] on sephadex G-50 column
spectrophotometry assay at wave length 254 nm

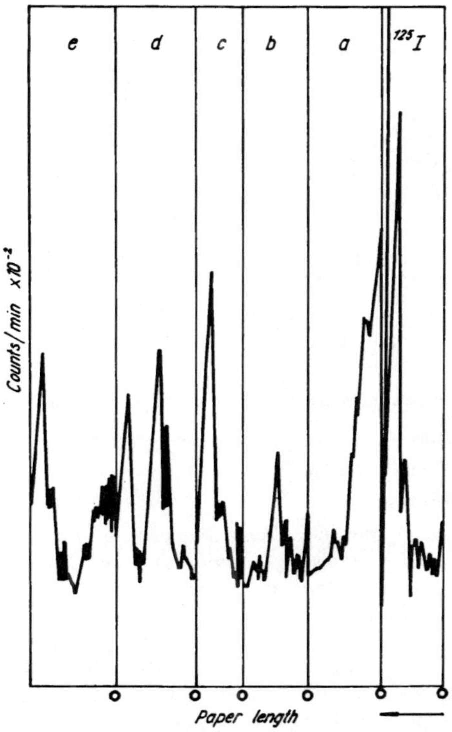

Fig. 3. Paper radiochromatography scanning for the separated peaks of ^{125}I-labelling of the peptide fragment [insulin (B_{3-6})]
a = Peak No. 1; b = Peak No. 2; c = Peak No. 3; d = Peak No. 4; e = Peak No. 5

Results

The fragment sequence showed that the position B_5, which is histidine residue is the only possible position to be iodinated [7]. The results obtained for identification of the fragment B_{3-6} are represented in Tab. 1 and Fig. 1–4.

Fig. 1 Radioactivity assay for purification of the ^{125}I-labelled peptide fragment (insulin B_{3-6}) on Sephadex G-50 column. Fig. 2 spectrophotometry assay for purification of ^{125}I-labelled peptide fragment (insulin B_{3-6}) on sephadex G-50. Fig. 3 paper radiochromatographic Scanning for the separated peaks of the labelled peptide fragment insulin B_{3-6}.

Tab. 1. RF Values of the fraction of ^{125}I-insulin (B_{3-6}) chain in comparison with the original amino acid and purified preparation using BAW system

Spotted	RF values	
	Colour	Radioscaning
First peak of ^{125}I peptide from		
Column chromatography	12	11
Second peak	51	51
Third peak	86	86
Fourth peak	49	50
Fifth peak	86	87
^{125}I as Na 1 Carrier free used in labelling	90	40
Insulin peptide fragment B_{3-6} prepared	86	
Histidine	13	—
Glutamine	21	
Asparagine	15	—
Leucine	64	—

BAW = Butanol: acetic acid: H_2O — 4:1:5

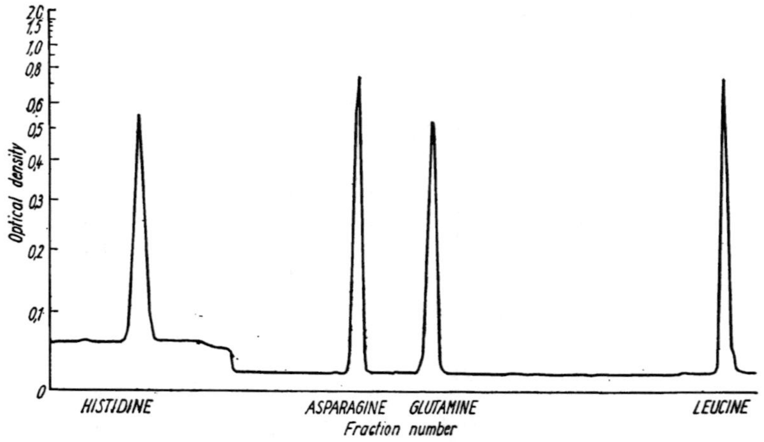

Fig. 4. Amino acid analysis for the synthesized peptide fragment [insulin (B_{3-6})]

Fig. 4 Amino acid analysis for the synthesized — peptide fragment (insulin B_{3-6}).

Discussion

The solid phase method of peptide synthesis has made possible the rapid preparation of wide variety of A and B [9] chain of insulin fragments as mentenod by *Merrifield* 1966. In our preparation of Leu-His Glu-Asp fragment of B-chain of insulin are concentrated on some specific non-protecting group, like that used by *Merrifield* [10] before, where it is the most easy method to introduce the amino acid and to remove the proteting groups without any side effect during the deprotection step. (Ztf) group was used for histidine amino acid as it is more stable than (Z) group which can not protect slightly with repeating coupling and deprotecting of-amino group with HCl/ACOH. (Mbh) Group 4·4 dimethoxy benzhydryl was used for protection of carboxyamides of Asp and Glu as it is the most reliable. for side chain protection of such type of amino acid.

Coupling of amino acids in the peptide synthesis has been carried out using dichlorohexylcarbodimide (DCC) as a tool. of cleavage for about 24 h at room temp. as it is the most popular and easy without any side effect and easy to remove on simple washing [11] for cleavage we used transestrification method to have the finished product in methyl-ester form which suppose to increase the biological activity of the product with less racemization.

Radioiodination of the peptide has been carried out using ^{125}I carier-free and chloramine Tas described by *Hunter* [12], with some modification to get high yield of specific activity and purity. For the identification of the prepared labelled peptide two tools are taken in consideration the first was the use of column chromatography using Sephadex G-50 and radioactivity assay of the eluted fractions which gave rise to five peaks (Fig. 1) from which the 3rd peaks was confirmed to be our purified insulin B_{3-6} labelled peptide. This is also confirmed with the spectrophotometry analysis (Fig. 2) for more confirmation paper radiochromatography was applied for the fractionated peaks and localisation of the compounds carried out by colour reaction (ninhydrine) and radioscanning using automatic paper radioscanning Tab. 1 (*Fresek* und *Hophner*) Fig. 3. In the mean time to conferm these results amino acid analysis has been carried out for our preparation Fig. 4.

Received December 15, 1982

References

[1] *M. A. Zewail*, M.R.C. Bull., Egypt **3** (1978) 217
[2] *H. Kappler, R. Schwyzer*, Helv. chem. Acta **44** (1961) 1136
[3] *J. P. Greenstein, M. W. Winitz*, chemistry of amino acids, Text book. Vol. 2, John Wiley and Sons Inc. New York 1961
[4] *R. B. Merrifield*, J. Am. Chem. Soc. **85** (1963) 2149
[5] *R. B. Merrifield*. Biochemistry Z. **1964**, 1385
[6] *J. M. Stewart, J. D. Young*, Solid phase peptide synthesis, Text book P 45 (1969)
[7] *C. H. Lie*, J. Am. Chem. Soc. **66** (1944) 225
[8] *M. F. Abdel-Wahab, F. X. Sulomo, S. Djajusman, N. Bambang, W. Natawiria*, Isotopenpraxis **18** (1982) 212
[9] *R. B. Merrifield, A. Marglin*, J. Am. Chem. Soc. **88** (1966) 5051
[10] *R. B. Merrifield*, J. Am. Chem. Soc. **86** (1964) 304
[11] *U. Weber, M. Andre*, Physiol.-Chem. **356** (1975) 701
[12] *W. M. Hunter, F. C. Greenwood*, Biochem. J. **89** (1963) 114

Isotopenpraxis **19** (1983) 8, S. 276–282

Informationen

IAEA. Heute gibt es auf der ganzen Welt ungefähr 300 Elektronenstrahlengeräte, und 110 Kobalt-60-Strahlenquellen sind in Betrieb. Der Wert der Produkte, die durch Strahlenverfahren hergestellt werden, beläuft sich auf 2–3 Milliarden US Dollar pro Jahr, wobei die Verkaufszahlen für solche Produkte im Laufe des vergangenen Jahrzehnts eine jährliche Wachstumsrate von 15 bis 20% verzeichneten. Man erwartet übrigens, daß sich diese sehr hohe Wachstumsrate auch in Zukunft halten wird.

Die großen Vorteile bei der Behandlung von Produkten mit Strahlen liegen in der Einfachheit der Technologie und in der dadurch erzielten Energieeinsparung, wie auf der 4. Internationalen Tagung über Strahlenverfahren in Dubrovnik, Jugoslawien, berichtet wurde. Über 270 Wissenschaftler und Ingenieure aus 32 Ländern trafen sich hier bei dieser von einem internationalen Sonderausschuß in Zusammenarbeit mit der IAEA vom 4. bis 8. Oktober 1982 veranstalteten Tagung, um Fortschritte bei der praktischen Anwendung von Strahlentechnologie und künftige Tendenzen auf diesem Gebiet zu besprechen.

In den Industrieländern ist eine Reihe von industriellen Strahlenverfahren eingeführt worden. Dies gilt vor allem für jene Industriezweige, die zum Beispiel Kunststoffe, Papier, Fahrzeuge, Gummi, Holz, Draht, medizinische Produkte und dergleichen herstellen oder verarbeiten, sowie für die Lebensmittelverpackungsindustrie. Durch Strahleneinwirkung können gewisse nützliche chemische Reaktionen hervorgerufen werden. Strahlen können zum Beispiel bei der Herstellung von Kunststoffen oder bei der Verbindung von Plastik-Kunststoffen mit anderen Materialien verwendet werden. Elektronenstrahlen finden in der Automobilindustrie bei der Behandlung von Belägen für Fahrzeugteile verbreitet Verwendung. Strahlenvulkanisierte Produkte wie Blattgummi und durch Bestrahlung querverbundene Schaumkunststoffe (zum Beispiel Polyäthylenfolien) werden als Aufprallschutz, als Futterstoff für Sturzhelme, Fußbodenmatten, für verschiedene Camping-Artikel und dergleichen eingesetzt. Abnützungsbeständige Holzprodukte können durch Bestrahlung von Holz-Kunststoff-Verbindungen hergestellt werden.

Damit ist die Reihe praktischer Anwendungsmöglichkeiten aber noch nicht erschöpft: Durch Bestrahlung von Lebensmitteln können beträchtlich längere Lagerzeiten erzielt werden, und auf dem Gebiet des Umweltschutzes gibt es auch neue Verfahren, zum Beispiel zur Desinfizierung von Klärschlamm und zur Entfernung von Schwefeldioxid und Stickstoffoxiden aus Abgasen. In Zusammenarbeit mit der IAEA werden jetzt besondere Bemühungen unternommen, die Strahlentechnologie an die Entwicklungsländer weiterzugeben, wobei die Anwendung dieser Technologie auf dem Gebiet der Sterilisierung von medizinischen Instrumenten und der Lebensmittelkonservierung von besonderer Bedeutung ist. (IAEA-Pressekommuniqué)

IAEA. Die Entwicklung von Leistungsreaktoren ist bisher vor allem auf der Grundlage der wassergekühlten Reaktoren erfolgt. Von den 281 Kernkraftwerken, die im September 1982 in 24 Ländern der Welt in Betrieb waren, arbeiten rund 230 mit wassergekühlten Reaktoren. Dadurch, daß bei diesen Reaktoren Wasser sowohl als Moderator als auch als Kühlmittel verwendet wird, kommt es bei den Brennstoffhülsen, beim Kühlsystem und bei anderen Bestandteilen zu Korrosion, Erosion und Hydration, wodurch die Sicherheit und Zuverlässigkeit des Kraftwerks beeinträchtigt werden kann. Durch Wanderung und Transport von Korrosionsprodukten und anderen Verunreinigungen bilden sich in manchen Teilen des Primärkreislaufs radioaktive Ablagerungen, die die Instandhaltung und Reparatur der Anlage erschweren.

Problemen dieser Art wird große Aufmerksamkeit zugewandt, mit dem Ergebnis, daß die Kernkraftwerke in insgesamt 2700 Reaktorjahren einen außergewöhnlich hohen Stand der Betriebssicherheit erreicht haben. Seit der Inbetriebnahme der ersten Anlagen in der Mitte der fünfziger Jahre ist es nicht zu einem einzigen Todesfall gekommen, der auf den Betrieb eines Kernkraftwerks oder seine Umweltwirkungen zurückzuführen gewesen wäre. Zur Erörterung dieses Themas veranstaltete die IAEA im November 1982 in Wien ein Internationales Symposium über die Chemie des Wassers und Korrosionsprobleme bei Kernreaktoren und ihren Bestandteilen. Bei dieser Gelegenheit wurden unter anderem die Auswahl von Materialien für Reaktoren und ihre Bestandteile, sowie ihre Konstruktion und Korrosionsbeständigkeit behandelt.

Auf dem Programm des Symposiums standen folgende Hauptthemen: bisherige Erfahrungen mit der Chemie des Wassers bei Kernkraftwerken mit verschiedenen Reaktortypen; Korrosion und Erosion von Baustoffen; einschließlich der Materialien für Brennelemente bei verschiedenen Reaktortypen; Errungenschaften der Dekontaminationstechnik für die Reaktoren von Kernkraftwerken; chemische und radiochemische Kontrolle von Reaktoren; technische Neuerungen bei der Reinigung des Kühlmittels im Primär- und Sekundärkreislauf von Reaktoren und neue Erkenntnisse bei den theoretischen Grundlagen und Modellen für die Untersuchung des Wasserhaushalts von Reaktoren und der Korrosion von Reaktorbaustoffen, einschließlich der Materialien für Brennstoffhülsen. (IAEA-Presseinformation)

IAEA. Die Auswahl eines Standortes für ein Kernkraftwerk ist ein komplexer Prozeß, bei dem man trachtet, das Risiko für die Bevölkerung, die Umweltwirkungen und die Kosten auf ein Mindestmaß herabzusetzen. Viele technische Disziplinen sind an der Wahl des Standortes für ein Kernkraftwerk beteiligt. Die wichtigsten davon sind Demographie, Geologie, Seismologie, Meteorologie und Hydrologie. Der Kraftwerksstandort wird nach überaus strengen Normen ausgewählt, um Gesundheit und Sicherheit der Bevölkerung zu gewährleisten.

Zur Erörterung der Fragen, die bei der Standortwahl in Entwicklungs- und Industrieländern auftreten, veranstaltete die IAEA im November 1982 in Wien ein internationales Seminar über Sicherheitsfragen bei der Standortwahl für Kernkraftwerke.

Im Jahre 1974 hat die IAEA mit der Durchführung eines Programms zur Erstellung von Sicherheitsnormen für Kernkraftwerke begonnen, das jetzt nahezu abgeschlossen ist. Eines der fünf Gebiete, die dabei behandelt werden, ist die Standortwahl. Für jeden Bereich wurden eine Sammlung von Richtlinien und mehrere Broschüren mit praktischen Anleitungen ausgearbeitet. Die Anweisungen für die Wahl von Standorten für Kernkraftwerke erläutern Methoden, mit deren Hilfe man ein größeres Gebiet zur Auswahl der in Frage kommenden Zonen untersuchen, die geeigneten Standorte innerhalb dieser Zonen und schließlich den geeignetsten dieser Standorte auswählen kann.

Sechs dieser Publikationen geben Anleitungen für den Schutz eines Kernkraftwerks gegen Naturkatastrophen: Erdbeben, Überschwemmungen und extreme klimatische Erscheinungen. Eine davon behandelt außergewöhnliche Ereignisse, wie Flugzeugabstürze, und benachbarte Industrieanlagen, die die Wahl des Standortes beeinflussen. Wenn das NUSS-Programm in ungefähr zwei Jahren abgeschlossen ist, wird es die erste international ausgearbeitete Normensammlung für einen so umfangreichen Bereich sein.

(IAEA-Presseinformation)

IAEA. Aus dem Inhalt des IAEA-Bulletins 24, No. 2, June 1982:

Nuclear power, man, and the environment; by *A. Hagen*

Developing guidelines on managing radioactive waste; by *D. Richter* and *S. Fareeduddin*

Nuclear waste management in Switzerland; by *R. Rometsch*

The back-end of the Swedish nuclear fuel cycle; by *G. Schultz*

A UK view of the magnaement of high-level waste; by *W. Marshall*

Why not in the ocean? by *C. L. Osterberg*

Natural and man-made radionuclides in the global atmosphere; by *Z. Jaworowski*

New directions in nuclear energy; by *G. Tavernier*

Safeguards — a model for general arms control? by *D. A. V. Fischer*
An overview of nuclear-weapon-free zones; by *G. Delcoigne*
The nuclear-weapon-free zone in Latin America; by *J. R. Martínez Cobo*
(IAEA-Bulletin)

IAEA. Aus dem Inhalt des IAEA-Bulletins 24, No. 3, Sept. 1982:
An assessment of nuclear energy in developing countries: how the Agency can help; by *L. L. Bennett, J.-P. Charpentier*, and *J. A. Marques de Souza*
Isotope hydrology techniques — practical tools to solve water problems; by *B. R. Payne*
Industrial technology transfer in Asia and the Pacific; by *E. E. Fowler*
Applying nuclear techniques to agriculture in Brazil; by *P. Willstätter-Greig*
International dose-assurance in radiation technology; by *K. H. Chadwick*
Predicting the behaviour of nuclear materials; by *P. Bull, J. D. Navratil, F. L. Oetting*, and *P. A. G. O'Hare*
A lawyer's view of safeguards and non-proliferation; by *L. W. Herron*
Conference report: Benefits and costs of pesticide use.

OECD-NEA. Die Nuclear Energy Agency (NEA) der OECD-Länder hat den 10. Aktivitätenbericht 1982 herausgegeben, in dem die Leistungen des Jahres 1981 enthalten sind. Der Bericht umfaßt 52 Seiten und ist in 8 Teilabschnitte gegliedert: Trends in der Kernenergie; Radiologische und Umwelteinflüsse durch Kernbrennstoffzyklen; Forschung zur nuklearen Sicherheit; Lizenzvergaben; Nukleare Entwicklung und Studien zum Kernbrennstoffzyklus; Gemeinsame Projekte; Nukleare Wissenschaften und schließlich Organisation und Administration. Zu den bemerkenswerten Trends des Jahres 1981 gehört die deutliche Steigerung der Kernkraftwerkskapazitäten in Frankreich (von 14 GWe Ende 1980 auf 22 GWe Ende 1981) und die Bemühungen in den USA und in der BRD, die Lizenzvergabezeitspannen wesentlich zu verkürzen.

Die Schrift ist von der OECD-NEA, 38, Bd Suchet 75016, Paris, Frankreich, herausgegeben worden.

Welt. Für die Kernkraftwerke der westlichen Welt wurden 1982 rund 33000 t Natur-Uran benötigt. Für das Jahr 1995 errechnete das Uranium Institute in London nach den derzeitigen Ausbauplänen für Kernkraftwerke in der westlichen Welt einen Bedarf von etwa 70000 t Natur-Uran. Der nur zur Stromerzeugung nutzbare Rohstoff Uran lieferte 1982 eine Energieausbeute, die etwa 360 Mio t Öleinheiten entspricht. Der Wert für 1995 entspräche einem Energieangebot von 770 Mio. t Öleinheiten für die Stromerzeugung. 1981 betrug der gesamte Ölverbrauch der USA 760 Mio. t Öleinheiten. Eine Tonne Öleinheiten hat den Heizwert von einer Tonne Rohöl.

(Atomwirtschaft/Atomtechnik 11/82)

Schweiz. Im September 1982 ist der 6. Bericht der Eidgen. Kommission für Strahlenschutz für das Jahr 1981 über die Dosimetrie der beruflich strahlenexponierten Personen in der Schweiz erschienen. Neben einer Übersicht der gesetzlichen Grundlagen und der verwendeten Methoden geht der Bericht zunächst auf eine Vergleichsmessung zwischen den 13 Dosimetriestellen ein, welche zum großen Teil innerhalb weniger Stunden Resultate im zugelassenen Streubereich abliefern konnten. Im Berichtsjahr wurden im ganzen bei 43431 Personen — 3% mehr als im Vorjahr — die Ganzkörperdosis durch äußere Bestrahlung gemessen. Die Kollektivdosis nahm gegenüber 1980 um 4,6% ab und erreichte noch 2062,1 manrem. Somit sank die mittlere Ganzkörperdosis auf 47 mrem. In Arztpraxen erreichte sie nur 24,4 mrem, in Spitälern lag sie bei 406,7 mrem und in Kernkraftwerken bei 913,1 mrem. Mit einer Ausnahme, einem Fall in einem Spital, lag die Ganzkörperdosis der Einzelpersonen unterhalb des jährlichen Grenzwerts von 5000 mrem. Die Überschreitung betraf eine Pflegeperson, bei der 5360 mrem gemessen wurden; sie hatte die Dosis bei der Pflege von Patientinnen mit Radium-Einlagen erhalten. Inzwischen sind Vorkehren zur Verhinderung solcher Fälle getroffen worden. Etwas überraschend waren ferner die vereinzelt hohen Dosiswerte in Zahnarztpraxen, wo bei normalem Verhalten nur minimale Dosen zu erwarten sind. Erstmals führt der Bericht die akkumulierten Gesamtdosen über den ganzen Zeitraum der beruflichen Tätigkeit, die sogenannte Lebensalterdosis, auf. Demnach gibt es in der Schweiz 142 erfaßte Personen, die eine Lebensalterdosis von mehr als 20000 mrem akkumuliert haben.

(SVA-Bulletin 17/82)

BRD/Großbritannien. Die britische nationale Strahlenschutzbehörde (National Radiological Protection Board) hat im Juni 1982 mit einer Untersuchung der Strahlenbelastung durch Baumaterialien in Wohnungen begonnen, die die Dosis-Messung in 2000 Wohnungen während einer Untersuchungsdauer von zwei Jahren vorsieht. Diese Strahlenbelastung der Menschen, eine Zusatzbelastung zur sog. natürlichen Strahlenbelastung im Freien, wurde auch in der Bundesrepublik Deutschland im Auftrag des Bundesministers des Innern von acht wissenschaftlichen Institutionen untersucht.

Über die Ergebnisse der rd. 30000 Messungen in Wohnungen und rd. 25000 Messungen im Freien in der BRD wird nachstehend kurz berichtet. Zum Vergleich: Die jährliche Abgabe an Radioaktivität durch kerntechnische Anlagen liegt nach dem Umwelt-Bericht des Bundesinnenministers unterhalb von 1 mrem.

Unter der Annahme eines Daueraufenthaltes im Freien würde die mittlere jährliche Keimdrüsendosis der Bevölkerung durch terrestrische Strahlung (Umgebungsstrahlung ohne die sog. Höhenstrahlung) etwa 38 mrem betragen. Unter der Annahme eines Daueraufenthaltes in Wohnungen würde die mittlere jährliche Keimdrüsendosis der Bevölkerung durch terrestrische Strahlung einschließlich der durch die Radioaktivität in Baustoffen bewirkten Strahlung etwa 50 mrem betragen. Die Höhe der jährlichen Keimdrüsendosis durch terrestrische Strahlung im Freien zeigt ausgeprägte regionale Unterschiede. Die in der Nähe von Wohngebieten festgestellten Werte liegen zwischen 3 mrem und 253 mrem. Analysen der spezifischen Aktivität verschiedener Materialien haben gezeigt, daß bei Granit Bims und anderem vulkanischen Gestein sowie bestimmten in größeren Mengen anfallenden Nebenprodukten industrieller Prozesse wie Schlacke und Schlämme überdurchschnittlich hohe Werte gefunden werden können. Unter ungünstigen Umständen kann bei ausschließlicher Verwendung eines Baustoffes die Differenz zwischen Aufenthalt im Freien und in Wohnungen bis zu 130 mrem betragen. Dies ist jedoch nur bei weniger als 1% der Bevölkerung der Fall. Die Ergebnisse schließen nicht die spezifischen Lungendosen ein, die zusätzlich durch radioaktive

Edelgase bewirkt werden, die im Freien und in Baumaterialien gebildet werden und in den Wohnungen in einer höheren Konzentration als im Freien mit der Luft inhaliert werden. Bei mangelnder Lüftung kann nach verschiedenen Untersuchungen sich daraus eine nennenswerte Lungenbelastung ergeben. (Atomwirtschaft/Atomtechnik 11/82)

Großbritannien. Die United Kingdom Atomic Energy Authority (UKAEA) hat bereits mehr als eine halbe Tonne Plutonium aus hochbestrahltem Brennstoff des 250-MW-Prototyp-Brüterkraftwerks PFR in Dounreay zurückgewonnen. Die Erfahrungen beim Betrieb der entsprechenden Wiederaufarbeitungsanlage waren außerordentlich günstig. Dies gab die UKAEA an der Internationalen Konferenz der IAEA über Erfahrungen mit Kernenergie im September 1982 in Wien bekannt. Ein Teil des Plutoniums ist inzwischen bereits zur Herstellung neuen Brennstoffs für den PFR verwendet worden. Dabei sind bei der Wiederaufarbeitung 0,44% und bei der anschließenden Brennstofffabrikation rund 0,9% des Plutoniums in den Rückständen und Anlagen zurückgeblieben. Durch Abfallbehandlung und Rezyklierung konnte laut UKAEA später der größte Teil dieses Materials wiedergewonnen werden. Die UKAEA hält aufgrund ihrer Erfahrungen die Machbarkeit der Schließung des Brüter-Brennstoffkreislaufs für erwiesen und plant nun die Erstellung einer Wiederaufarbeitungsanlage mit einer Kapazität von 50 Jahrestonnen. Dabei liegt der Entwicklungsschwerpunkt bei der Verbesserung der mechanischen Komponenten und nicht mehr bei der chemischen Verfahrenstechnik. Bereits in der bestehenden Anlage erfolgte der Aufschluß der Brennelemente mit einem Laser anstelle einer Schlagschere. (SVA-Bulletin 17/82)

Großbritannien. Von der United Kingdom Atomic Energy Authority (UKAEA) wurden folgende Reports herausgegeben:

AERE-R 10154 The formation, identification and elimination of various ionic species of iodine occurring in the production of iodine-123. By *R. Clegg*, *A. Roberts*, *H. E. Sims*, and *J. A. Winter*, June 1982. 18 pp.

AERE-R 10485 Radioactive fallout in air and rain: Results to end of 1981. By *R. S. Cambray*, *K. Playford*, and *G. N. J. Lewis*. July 1982. 43 pp.

AERE-R 10493 Experiments with higher-order finite difference formulae. By *C. P. Thompson* and *N. S. Wilkes*. July 1982. 59 pp.

AERE-R 10552 Notes on coherent Raman scattering. By *P. Hutchinson*. June 1982. 32 pp.

AERE-R 10523 Small angle neutron scattering (SANS) from heat treated PE16 specimens from Berkeley Nuclear Laboratories. By *C. G. Windsor* and *V. S. Rainey*, June 1982. 27 pp.

AERE-R 10553 Notes on signal processing in laser Doppler anemometry. By *P. Hutchinson*. June 1982. 18 pp.

AERE-R 10567 The sources and fates of atmospheric sulphur compounds in the United Kingdom. By *R. G. Derwent*. June 1982. 33 pp.

AERE-R 10568 Functional decomposition of real-time systems. By *M. H. Gilbert*. June 1982. 15 pp.

AERE-R 10608 Finite element solutions for flow in a symmetric channel with a smooth expansion. By *K. A. Cliffe*, *C. P. Jackson*, and *A. C. Greenfield*. August 1982. 54 pp.

CLM-R 221 The physics of magnetic confinement configurations: Tokamak theory and experiment. By *D. C. Robinson*. January 1982. 32 pp.

CLM-R 223 Deuterium trapping in low Z coatings. By *S. K. Erents*. April 1982. 94 pp.

ND-R 751(S) Surfaces, coatings and protection. UKAEA diffraction analysis conference, 29th meeting, Sellafield, 20—22 October, 1981. Edited by *I. F. Ferguson*. August 1982. 122 pp.

ND-R 755(R) Mesh-size errors in diffusion-theory calculations using finite-difference and finite-element methods. By *A. R. Baker*. July 1982.

ND-R 816(R) Photoelastic investigation of the stress concentration factor in welded joints. By *P. S. Wardle*, *R. Chegwin*, and *K. Reading*. July 1982. 61 pp. (Atom 313/82)

USA. In der neuesten Ausgabe seiner „Enrichment News" (September 1982) legt das amerikanische Department of Energy (DOE) seinen Operationsplan für die Urananreicherung in einer kurzen Zusammenfassung dar. Im Vordergrund steht dabei die teilweise kontroverse Frage der Einführung der Urananreicherung mittels Gaszentrifugen. Ein Vergleich der vom DOE untersuchten Optionen für die 20-Jahres-Periode 1983 bis 2002 zeigt, daß die Fortsetzung des Baus der Gas Centrifuge Enrichment Plant (GCEP) bei Portsmouth, Ohio, höchst wünschenswert ist. Gemäß jetzigem Zeitplan sollen dort die ersten zwei Gebäude bis 1989 fertiggestellt werden, während weitere Gebäude zwischen 1990 und 1994 betriebsbereit werden sollen. Dank der GCEP werde der Preis für Anreicherungsdienste langfristig wesentlich reduziert werden, schreibt das DOE. Den Kunden brächte die Zentrifugenanreicherung des DOE Einsparungen von $ 5,5 Mia. Die Einsparung elektrischer Energie wird mit 29000 Megawatt-Jahren angegeben, was der totalen Produktion zweier großer Kraftwerke während 20 Jahren entspreche. (SVA-Bulletin 19/82)

Namibia. Die Rössing-Uranmine in Namibia, zu 46,5% im Besitz der britischen Rio-Tinto-Zinc-Gruppe (RTZ), kann trotz zurückgehender Uranpreise auf dem Weltmarkt noch ansehnliche Gewinne vorweisen. Diese größte Uranmine der Welt, die 1976 die Produktion aufgenommen hatte, steuerte in der ersten Hälfte dieses Jahres mit £ 13,5 Mio einen ansehnlichen Betrag zum Gesamtnettogewinn der RTZ bei.

Von Problemen bleibt die Uranmine indessen nicht verschont. Auf dem freien Markt ist der Uranoxidpreis pro Pfund auf unter US $ 17 gesunken. Da die Urankonzentration in Rössing relativ gering ist, müssen sehr große Mengen gefördert werden, damit eine gewinnbringende Produktion möglich ist. Die Auslastung liegt heute bei 93%, und die Produktion beläuft sich auf rund 4500 Tonnen. Ein großer Teil des Absatzes ist durch längerfristige Verträge abgedeckt. Die Tatsache, daß Rössing bisher keine Steuern bezahlen mußte, stellte eine zusätzliche Erleichterung dar.

Neben den Unsicherheiten des Uranmarktes kennt die namibische Uranmine zusätzlich noch politische Schwierigkeiten. Sie rechnet fest mit der kommenden Unabhängigkeit Namibias und versucht sich entsprechend darauf vorzubereiten. Erziehungs-, Gesundheits- und Sozialprogramme sollen korrekte Beziehungen mit einem souveränen Staat garantieren. Auch werden die Löhne der Arbeiter nicht nach rassistischen Kriterien berechnet. (SVA-Bulletin 18/82)

Australien. Am 12. Oktober 1982 wurde die Produktion von Uranoxid in Australiens ältester Uranmine eingestellt. Die Produktion wurde im Sommer 1958 aufgenommen, und Mary Kathleen blieb während Jahren die einzige Uranmine Australiens. Der Abbau der verbleibenden Reserven ließ sich nun ökonomisch nicht mehr rechtfertigen. Darüber hinaus besitzt die Rio-Tinto-Zinc-Gruppe genügend Uran, um den laufenden Verpflichtungen nachkommen zu können. (SVA-Bulletin 18/82)

Australien. Am 22. September 1982 äußerte sich der australische Vize-Premierminister und Minister für Handel und Bodenschätze *Anthony* im nationalen Parlament über den Stand und die Entwicklungsmöglichkeiten der australischen Uranindustrie. Mit Befriedigung unterstrich er, daß das Land eine gesunde, profitable und expandierende Uranindustrie besitzt. Er betonte, daß Produktion, Exporte und Beschäftigung auf Rekordhöhen geklettert sind. Ständig werden überdies neue Uranvorkommen erschlossen.

Gemäß Anthonys Rede sind Australiens bekannte Uranreserven auf einem Rekordstand. Sie belaufen sich auf ungefähr 294000 t (ohne Roxby Downs) und machen 19% der westlichen Reserven an Uran aus, das für weniger als US $ 80 pro kg abgebaut werden kann. Während 1981 nur 3300 t U_3O_8 produziert wurden, dürfte die Produktion im Jahre 1982 5000 t U_3O_8 überschreiten. Dementsprechend entwickelten sich die Exporte. (SVA-Bulletin 19/82)

DDR. Der Leipziger Kommissions- und Großbuchhandel (LKG) hat den Literaturkatalog 1982 „Elektrotechnik—Elektronik—Regelungstechnik" herausgegeben (Redaktionsschluß 3. 9. 1982). Er ist mit einem Titel- und Autorenverzeichnis versehen und umfaßt 47 Seiten. Der Katalo enthält die zum genannten Themenkreis erschienenen ung durch den LKG lieferbaren Titel der DDR-Verlagsprodukd tionen.

BRD. Im November-Heft 1982 der Zeitschrift Atomwirtschaft/Atomtechnik sind folgende Artikel mit statistischen Angaben enthalten:

— Ein- und Ausfuhr der BRD an Kernbrennstoffen 1981
— Der Außenhandel der BRD mit kerntechnischen Erzeugnissen 1979—1981
— Ein- und Ausfuhr der BRD an Radionukliden 1981
— „Besondere Vorkommnisse" in den Kernkraftwerken der BRD 1981 (Atomwirtschaft/Atomtechnik 11/82)

BRD. Die vom Verlag Chemie, Weinheim (Postfach 1260/1280) herausgegebene Buchreihe „Vom Wasser" enthält folgende ausgewählte Beiträge, die im Zusammenhang mit radiochemischen oder isotopenhydrologischen Methoden stehen:

Bd. 56 (1981):
Isotopenhydrologie — Stand und Ausblick (*Moser*)
Zur Frage des Nitratgehaltes von Fließgewässern — dargestellt am Beispiel der Ruhr (*Kornatzki, Koppe*)
Bd. 55 (1980):
Radiochemische Untersuchungen zum Verhalten von Quecksilber- und Methylquecksilberchlorid in Süß- und Meerwasser (*May, Reisinger, Flucht, Stoeppler*)
Hochtemperaturfiltration in Kernkraftwerken zur Reduzierung der Korrosionsprodukte und der Radioaktivität (*Heitmann*)
Bd. 54 (1980):
Verhalten radioaktiver Verunreinigungen im Dampf-Wasserkreislauf eines Hochtemperatur-Reaktors (*Nieder, Vey*)
Abtrennung und energiedispersive RFA gelöster Schwermetallspuren, speziell in Trinkwässern, mittels chelatbildender Celluloseaustauscher (*Burba, Dyck, Lieser*)
Grundlagen der statistischen Behandlung einfacher linearer Korrelationen (*Wagner*)
Bd. 53 (1979):
Verbesserung chemischer und physikalischer Methoden zur Aufbereitung radioaktiver Abwässer aus Kernkraftwerken (*Oschmann, Schiffers*)
Bd. 52 (1979):
Vorrichtung zur registrierenden radiometrischen Messung der Konzentration von Feststoffen und Schwermetallen in Wasser (*Dresia, Brämer, Frank, Koenen, Toplek*)
Bd. 50 (1978):
Untersuchung von Industrieabfällen mit der energiedispersiven Röntgenfluoreszenz (*Hilger, Becker*)

IAEA/GSF. An agreement between the IAEA and the Gesellschaft für Strahlen- und Umweltforschung (GSF) on a joint research programme for the application of isotopic techniques in hydrological and geothermal investigations in Latin America was signed in October 1982.

The aim of the programme is the promotion of isotopic techniques, especially those based on environmental isotopes, in the development of groundwater resources for human and agricultural use and of geothermal resources for power production. Isotopic techniques are especially useful in investigations into the origin of groundwaters; the rate of groundwater recharge; the age or residence time of groundwater bodies; the movement of groundwater; the relationship among groundwater systems; and the temperature and availability of geothermal fluids.

The individual projects will be carried out largely by Latin American institutions in response to local conditions and requirements, with advice and cooperation from the IAEA and the GSF Institute of Radiohydrometry.

The new agreement, which will last until the end of 1986, follows another joint IAEA/GSF programme, which terminates in 1983. It consists of basic research into the mechanisms of groundwater recharge. (Atom 314/1982)

IAEA. Die IAEA hat im November 1982 eine 2. Ergänzung zu ihrem Veröffentlichungskatalog 1981/1982 herausgegeben:

IAEA publications Catalogue 1981/82, 2nd Supplement 18 pp. Die 1. Ergänzung ist im Januar 1982 erschienen.

DDR. Vom Staatlichen Amt für Atomsicherheit und Strahlenschutz (SAAS) der DDR sind folgende neue Berichte erschienen:

SAAS-290 Bestimmung der Ablagerungstiefe niederener getischer Quantenstrahler im Gewebe mittels hochauflösender Röntgenspektrometrie (Depth determination of low-energy photon emitter deposits in tissue by means of high-resolution X-ray spectrometry; *W. Schlüter*) 98 S.

SAAS-291 Stand und Entwicklungstendenzen des chemischen Strahlenschutzes — Literaturstudie 1979/1980 (State and tendencies of chemical protection against ionizing radiation — a literature survey 1979—1980; *G. Siegel, E. Tapp, H. Hannig, R. Dlaske, W. Papendieck, K. Martinek* u. *J. Hähn*) 146 S.

SAAS-293 Ein Verfahren zur Entfaltung hochenergetischer Szintillations-Gammaspektren bis 8 MeV (A method for unfolding high-energy scintillation gamma-ray spectra up to 8 MeV; *N. Dymke* u. *B. Hofmann*) 22 S.

SAAS-294 ^{85}Kr-Nachweis mittels *Cherenkov*-Strahlungsmessung (^{85}Kr detection by *Cherenkov* radiotion measurement; *H.-J. Scheibe*) 28 S.

SAAS-297 Vergleich von Prüfmethoden für Aerosolfilter (Comparison of testing methods for aerosol filters; *S. Przyborowski* u. *W. Ullmann*) 39 S.

Schweiz/USA. Die USA und die Schweiz unterzeichneten am 15. Nobember 1982 in Bern ein Abkommen über die Erweiterung ihrer Zusammenarbeit auf dem Gebiet der Reaktorbrennstoffentwicklung. Das Abkommen zwischen dem US Department of Energy (DOE) und dem Eidgenössischen Institut für Reaktorforschung (EIR) erstreckt sich über acht Jahre und umfaßt Arbeiten für Brüterbrennstoff, die am EIR in Würenlingen und im Los Alamos National Laboratory, USA, ausgeführt werden.

Die Zusammenarbeit zu diesem Entwicklungsprogramm wurde im April 1981 aufgenommen. Sie schließt die gemeinsame Bestrahlung eines ganzen Karbid-Brennstoffbündels ein und wird den Vergleich von Los Alamos-Brennstoff und schweizerischem Brennstoff ermöglichen. Karbidbrennstoff gestattet vor allem eine höhere Brennstoffausnützung im Vergleich zu dem heute gebräuchlichen Oxid-Brennstoff. Der schweizerische Fabrikationsprozeß verspricht zusätzliche Vorteile auf dem Gebiet der Wiederaufarbeitung und der anschließenden Herstellung von neuen Brennelementen. Die Brennstoffhüllrohre für das Testbündel werden in Los Alamos hergestellt, dann in der Schweiz mit Brennstoff gefüllt und nach strengen US-Vorschriften zugeschweißt. Die schweizerischen Brennstäbe werden zusammen mit Los Alamos-Stäben während zwei Jahren in der amerikanischen Fast Flux Test-Anlage in Hanford bestrahlt. Nach der Entnahme aus dem Reaktor werden die Elemente 18 Monate lang gekühlt und anschließend in Los Alamos durch Wissenschaftler beider Länder einem ausgedehnten Testprogramm unterzogen.

(SVA-Bulletin 20/1982)

Großbritannien. In letzter Zeit sind folgende Reports der United Kingdom Energy Authority (UKAEA) erschienen:

AERE-M 3252 A gamma doserate surveillance system. By *A. T. Rolls, G. F. Snelling* and *F. D. Seymour*. July 1982. 13 pp.

AERE-R 10032 Random number generators for the normal and gamma distributions using the ratio of uniforms method. By *I. Robertson* and *L. A. Walls*. December, 1980. 58 pp.

AERE-R 10470 Determination of rare earth elements in geological materials by neutron activation analysis using a group separation and high resolution gamma-ray spectrometry. By *T. W. Sanders* and *S. J. Wright*. September, 1982. 21 pp.

AERE-R 10554 The simultaneous measurement of particle size, velocity and mass transfer in a pulsed two-phase flow field. By *M. L. Yeoman, N. S. Lightfoot* and *A. P. Morse*. June, 1982. 21 pp.

AERE-M 3258 A beta-in-air monitor surveillance, alarm and record system. By *G. F. Snelling* and *F. D. Seymour*. August, 1982. 26 pp.

AERE-M 3261 A process plant cooling water monitor for fission product breakthrough. By *A. T. Rolls, G. F. Snelling* and *F. D. Seymour*. August, 1982. 17 pp.

AERE-R 9066 (rev) Initial experience of manufacture and property measurements on piezoelectric cable. By *B. H. Broomfield*. November, 1981. 23 pp.

AERE-R 10496 Selection of low-activity elements for inclusion in structural materials for fusion reactors. By *O. N. Jarvis*. June, 1982. 46 pp.

AERE-R 10631 Computer modelling studies of photochemical air pollution formation in power station plumes in the United Kingdom. By *R. W. Derwent*. August, 1982. 54 pp.

ND-R 461(S) The *Auger* spectra of some ternary lithium compounds. By *I. F. Ferguson, D. R. Masters* and *M. Turek*. September, 1982. 19 pp.

ND-R 803(W) A comparison of measured fission gas releases for Windscale AGR fuel irradiated above 18 GWd/tU with those predicted using the computer code MINIPAT D. By *N. Beatham, R. Hargreaves, R. D. Walker* and *R. Kirkbride*. August, 1982, 26 pp.

(Atom 314/1982)

Tunesien. Die von der französischen Uranium PUK (UPK) errichtete Pilotanlage für die Gewinnung von Uran aus Phosphatsäure bei Gabès in Tunesien wurde im November 1982 vom Staatspräsidenten *Bourguiba* eingeweiht. Die Pilotanlage ist der erste Schritt auf dem Weg zu einer Anlage, die die gesamte Phosphorsäureproduktion in Gabès ausbeuten soll und deren Baubeginn für Ende 1983 geplant ist. Auf Grund eines Lizenzvertrages der Industries chimiques maghrèbines (ICM) mit der UPK vom April 1982 wird in Gabès ein von der amerikanischen Gruppe Gardinier und von der UPK entwickeltes Verfahren zur Extraktion von Uran aus Phosphorsäure verwendet. Es wird erwartet, daß in Gabès so jährlich 120 t Uran als Nebenprodukt der Düngerproduktion gewonnen werden können.

(SVA-Bulletin 20/1982)

Japan. Der zuständige Electric Power Resources Development Coordination Council (EPRDCC) der japanischen Regierung genehmigte am 21. September 1982 unter dem Vorsitz des Premierministers die Pläne für die Nuklearblöcke Genkai-3 und -4 der Kyushu Electric Power Company. Beide Anlagen sollen mit Druckwasserreaktoren ausgestattet werden und eine Leistung von je 1180 MW aufweisen. Damit werden sie die größten nuklearen Einheiten Japans sein. Laut Angaben des Japan Atomic Industrial Forum sind in Japan jetzt 43 KKW-Blöcke in Betrieb, im Bau oder bewilligt.

Am gleichen Standort betreibt die Kyushu Electric Power Co. die beiden PWR-Einheiten Genkai-1 und -2 (je 529 MW). Diese stehen seit 1975 und 1981 in Betrieb. Genkai-1 verzeichnete am 22. Oktober 1982 einen Zuverlässigkeits-Rekord, als der Block nach 367 Tagen ununterbrochenem Vollbetrieb zum Brennstoffwechsel abgeschaltet wurde. (SVA-Bulletin 20/1982)

VR China. Für Chinas erstes 300-MW-Kernkraftwerk wurde kürzlich der Standort Ginshan, Haiyan County, in der Provinz Zhejiang, festgelegt. Ginshan liegt an der Hangzhou Bay, nicht weit von Shanghai, der größten Stadt Chinas. Der Standort Ginshan wurde am 11. November 1982 vom Minsterium für Nuklearindustrie bekanntgegeben. Das Ministerium bestätigte gleichzeitig, daß das erste Kernkraftwerk „made in China" mit einem Druckwasserreaktor ausgestattet werden soll. (SVA-Bulletin 20/1982)

UdSSR. In Heft 5/1982 der Zeitschrift „Atomnaja Energija", Jahrgang 53, sind u. a. folgende Beiträge *in russisch* enthalten:
25th Anniversary of the IAEA (*I. G. Morozov*)
Thermoradiolysis of Carbon Dioxide (*B. G. Dzantiyev, A. N. Ermakov, B. M. Zhitomirski, V. N. Popov*)
Increasing the Efficiency of Electron Beam Irradiation of Cylindrical Objects by Means of Magnetic Systems (*L. L. Akrachkova, N. K. Kuksanov, R. A. Salimov, Yu. N. Timko*)
Non-destructive Method of Determining the Plutonium/Uranium Ratio in Fast-Reactor Fuel-Elements by X-ray Spectrometry (*A. V. Bushuyev, V. I. Galkov, A. V. Zvonarev, A. F. Zolotov, A. A. Kutuzov, N. A. Melnichenko, V. N. Ozerkov, V. V. Chachin*)
Thermobarostable Gamma-Ray Spectrometer for Investigation of Ultra-deep Bore Holes (*E. M. Arm, A. A. Ilyinski, B. E. Metsger, V. I. Pyatakhin*)
On Distance Measurements of Ionizing Radiation Fields in Boreholes Using Optical Fiber Cable (*V. A. Andronov, V. I. Balayev, E. M. Dianov, A. A. Ilyinski, E. V. Karus, O. L. Kuznetsov, E. V. Mishin, L. G. Petrosyan, A. M. Prokhorov, V. I. Pyatakhin*).

BRD. In Heft 1/1983, der Zeitschrift „Atomwirtschaft/Atomtechnik", 28. Jahrgang, sind u. a. folgende Beiträge enthalten:
K. W. Deutsch, Gesellschaftliche Aspekte der Energieversorgung
H. Albrecht u. *H. Wild*, Experimentelle Bestimmung von Quelltermen beim LWR-Kernschmelzen
W. Seifritz u. *P. Wydler*, Stromgestehungskosten beim Übergang vom DWR zum FDWR (fortgeschrittener Druckwasserreaktor)
H. J. Laue, Stand der Entwicklung von Kernkraftwerken kleiner und mittlerer Leistung
Ch. Coninx, Öffentlichkeitsarbeit als Bürgerinformation
H. Trauboth u. *U. Voges*, Verfahren zur Entwicklung zuverlässiger Software für rechnergestützte Sicherheitssysteme
Tagungsbericht über Beschleunigeraspekte der schwerioneninduzierten Trägheitsfusion
Tagungsbericht über Fortschritte bei Konditionierung, Zwischen- und Endlagerung radioaktiver Abfälle.

DDR. Das Binnen- und Außenhandelsunternehmen Isocommerz GmbH hat folgende Kataloge, Prospekte und Tagungsmaterialien herausgegeben:

- Stable Isotopes Labelled Compounds, 38 p., September 1982
- Schriftenreihe „Anwendung von Isotopen und Kernstrahlen in Wissenschaft und Technik"
  - No. 5: ^{15}N-Anwendung in der Biochemie, Landwirtschaft und Medizin, Eine Einführung, Berlin 1981, 93 S.
  - No. 12: Neuronuklearpharmaka, Berlin 1982, 121 S.
- Schriftenreihe „Wissenschaftliche Tagungen in der DDR"
  - XVII. Nuklearmedizinisches Symposium, Reinhardsbrunn, vom 27.—31. Mai 1980, 415 S.
  - XVIII. Nuklearmedizinisches Symposium, Reinhardsbrunn, vom 7.—11. 12. 1981, 226 S.
  - I. Bilaterales Symposium DDR—ČSSR, Reinhardsbrunn, vom 18.—21. 5. 1981, 293 S.

Die aufgeführten Materialien können von der Isocommerz GmbH, DDR-1115 Berlin, Lindenberger Weg 70, angefordert werden.

Isotopenpraxis 19 (1983) 8, S. 282—292

Buchbesprechungen

Forscher an den Wurzeln des Seins
Die abenteuerliche Welt der Elementarteilchen

W. Braunbek u. K. Röttel. Düsseldorf — Wien: Econ Verlag 1981; 461 S., 93 Abb., 27 Photos, mehrere Tab.; Leinen DM 58,—

Wenn der moderne Mensch beginnt, über sich selbst, seine Umwelt und über die Funktionsweisen vieler ihn umgebender Dinge nachzudenken, so stehen ihm eigentlich nur wenige Möglichkeiten zur Verfügung, um die vielen selbstaufgeworfenen Fragen zu beantworten. Er kann mit Ernst und Fleiß ein Studium (oder Selbststudium) aufnehmen; er kann sich geduldig mit Fachleuten unterhalten — wobei auch die Geduld der Fachleute arg strapaziert werden kann —, oder er kann sich Bücher verschaffen, die ihm zumindest einen Einblick in ein bestimmtes Fachgebiet erlauben, ihm eine Ahnung davon vermitteln, worum es geht und warum diese oder jene Probleme gegenwärtig angegangen und bearbeitet werden und wo die Schwierigkeiten liegen. Aufregende Kapitel der Raumfahrt, der Gentechnologie, der Archäologie und der Physik werden eben geschrieben, und eines der besonderen Kapitel ist das der Elementarteilchenphysik, da die gigantischen Maschinen nicht nur beeindruckend und kostspielig sind (immerhin sollen die Baukosten vom Projekt HERA runde 650 Millionen DM betragen, was 1500 Eigenheimen entspricht), sondern häufig durch ihre Superlative die Aufmerksamkeit auf sich lenken.

Zwei Autoren — ein älterer, der 1977 in Tübingen verstorbene Professor für theoretische Physik *Braunbek*, und ein jüngerer, der 1939 geborene *Karl Röttel* — unternehmen den Versuch, die Geschichte der Physik der Elementarteilchen zu schreiben und dem Leser das Ringen um Erkenntnis vieler daran beteiligter Persönlichkeiten und Forschergruppen darzustellen. Die Geschichte wird in 50 Kapitel unterteilt, die leicht überschaubar sind. Die ersten 25 Kapitel (bis S. 132) reichen bis zur Unschärferelation *Heisenbergs*, die folgenden Kapitel (26 ... 48) sind der QED, den *Feynman*-Diagrammen, den Erhaltungssätzen und den Quarks gewidmet, also wenn man so will, dem gegenwärtigen Wissensstand auf diesem Gebiet.

Das Buch ist für Leser geschrieben, die über ein von einer mittleren Reife übriggebliebenes Wissen verfügen, wie es im Vorwort heißt. Es ist schwer, genau zu definieren, was damit an Wissen vorausgesetzt wird, aber die Exponentialschreibweise von Zahlen als Produkt aus einer Grundzahl (Mantisse) und einer Zehnerpotenz sollte man einfach als gegeben hinnehmen; in jeder Bedienungsanleitung eines Taschenrechners wird auch mit dieser Schreibweise gearbeitet, so daß eine nochmalige Einführung auf den Seiten 61—64 sich erübrigen sollte. Vorher wird mit den Zahlenwerten Trillionen und Sextillionstel u. a. m. operiert; so wird z.B. die Ruhemasse eines Protons als 1672,5 Quintillionstel Kilogramm angegeben, was keineswegs der Anschaulichkeit dient. Das zweite, was sich störend beim Lesen auswirkt, ist das Fehlen einer Numerierung der an sich guten 27 Photographien zwischen den Seiten 160 und 161; ein jeder Photohinweis im Text verlangt vom Leser ein Durchzählen dieser Photos; nebenbei gibt es kein o. l., o. und u. gegenüber der Seite 161. Ferner sollten die zu häufig eingeschobenen Hinweise der Art „... siehe späteren Abschnitt", „später hören werden", andernorts", „... in den nächsten Kapiteln ..." u. ä. stark eingeschränkt werden, da sie ohne nähere (genauere) Seitenangaben nur redundante Informationen sind. An kleineren Flüchtigkeits- bzw. Druckfehlern wären zu nennen: „nature" S. 171, ε^- in Tab. auf S. 220, $\left(\frac{5}{3}\right)$ auf S. 299 für die Kombinationen mit Wiederholung, hf auf S. 302, 20 K auf S. 310 und der Seitenhinweis im Inhaltsverzeichnis S. 446 für *H. Lindner*, 158ff. sollte S. 159 sein. Die zu kleinen Schriftgrade (Nonpareille bis Perl) bei Formeln sind schlecht lesbar, beispielsweise S. 198, 279, 282 u. 283. Die Tabelle auf S. 282 ist in ihren Spalten total verschoben, hier dürfte nur ein Querformat weiterhelfen. Das c im Nenner der Formel auf S. 309 ist überflüssig, und die Abszisse zu Abb. 44.1 auf S. 327 ist einerseits verschoben und sollte andererseits nicht in GeV, sondern in MeV eingeteilt sein.

Einige Richtigstellungen sollten bei künftigen Neuauflagen bedacht werden: Auf S. 51 wird *Maxwell* als Gegner der Quantentheorie genannt — er ist aber 21 Jahre vor *Plancks* Entdeckung bereits gestorben. Der berühmte Streit zwischen *Rutherford* und *Ramsay* um die 20 mg Ra auf S. 36/37 ist zu knapp und zu ungenau wiedergegeben. Im Bastnäsit (S. 183) wird primordiales ^{244}Pu nachgewiesen, aber nicht ^{237}Np und ^{239}Pu, die beide bereits in den Jahren 1948 ... 1952 in Uranmineralien gefunden wurden (*D. F. Peppard* et al und *G. T. Seaborg* et al.). Auf S. 186 werden die Namen der Transurane so diskutiert, als wäre dies ein Problem der Gewohnheit („... Ku scheint sich nicht durchgesetzt zu haben ...") — wohl kaum, hier sollte deutlich gemacht werden, daß unterschiedliche Kriterien der Anerkennung von Nachweisen künstlicher Elemente die eigentliche Ursache sind, wie überhaupt die schönen Ergebnisse des SHIP, des Separator for Heavy Ion Reaction Products, vom GSI Darmstadt (Prof. *P. Armbruster* und Dr. *G. Münzenberg*) gar nicht mit genannt werden. Daß heute die schwersten Elemente ^{262}107 (25. 2. 1981 — 1.^{30} nachts, aus $^{54}_{24}$Cr auf $^{209}_{83}$Bi-Target) und 109 (Sept. 1982, aus $^{58}_{26}$Fe auf $^{209}_{83}$Bi-Target) sind, deutet auf die schnelle Entwicklung gerade auf diesem Gebiet hin. Das sowjetische Team um *Ju. Z. Oganesyan, Demin* et al. in Dubna hat das Nuklid ^{261}107 1976 erzeugt.

Das Buch erfüllt in weiten Bereichen das Informationsbedürfnis des vielzitierten interessierten Laien und bringt ihm die zunächst fremde Welt der hohen und höchsten Energien zur Suche der kleinsten Teilchen in der Natur näher. Damit ist das Gesamtanliegen des Buches erreicht — die abenteuerliche Welt der Elementarteilchen aufregend zu erleben und — wenn möglich — auch etwas davon zu verstehen.

H. Hübner

Biophysik

W. Hoppe, W. Lohmann, H. Markl u. H. Ziegler (Hrsg.), 2. völlig neubearb. Aufl.; Berlin — Heidelberg — New York: Springer 1982; 980 S., 856 Abb., 99 Tab.; gebunden DM 168,—; US $ 74.60

Man könnte in Abwandlung eines arabischen Sprichwortes von der 2. Auflage der Biophysik sagen: Dem ersten — der 1. Auflage — gebührt der Ruhm, die weite Anerkennung in der Fachwelt, wenn auch die Nachfolger — die Neuauflage — es besser gemacht haben. Um aktuell zu sein und zu bleiben, haben sich die Herausgeber der im Vorwort angedeuteten Mühe unterzogen, eine Neuauflage erscheinen zu lassen, die man wegen ihrer sehr umfassenden Neubearbeitung fast als ein neues Werk auffassen kann. Durch die sinnvolle Erweiterung (Seitenumfang nimmt von 706 auf 963 S. — ohne Sachverzeichnis — zu; die Zahl der Autoren steigt von 52 auf 68) und Verbesserung wird eine Neufassung angeboten, die an Zuverlässigkeit der zusammengestellten verbindlichen Tatsachen über die physikalischen Aspekte der Lebensvorgänge gegenwärtig kaum übertroffen werden kann. Kaum eines der 18 Kapitel der 1. Auflage (in der neuen Auflage sind es übrigens nur 17 Kapitel, was auch auf eine starke Umarbeitung hindeutet) ist von einer Überarbeitung verschont geblieben. Von den neuaufgenommenen Themen wären zu nennen: diffuse Kleinwinkelstreuung von makromolekularen Lösungen, photoakustische Spektroskopie, Wirkungsspektrometrie (in der 1. Auflage noch Aktionsspektrometrie genannt), Konformationsumwandlungen von Biopolymeren, strukturelle Organisation der Proteine, Thermodynamik und Kinetik von Self-Assembly-Vorgängen, Membranpotentiale, Biolumineszenz, Biophysik der Fortbewegung auf dem Lande, Biophysik biologischer Oszillationen und schließlich die chemische Evolution und der Ursprung lebender Systeme. Man ist bei dieser Aufzählung geneigt, sich zu fragen, wie eine 1. Auflage ohne diese bedeutsamen Teilgebiete überhaupt auskommen konnte, oder sollte etwa durch den Reiz des Details die nach einer Verallgemeinerung strebende Aussage verlorengegangen sein? Mitnichten, denn gerade diese Detaildarstellungen entstammen der Feder sachkundiger Autoren.

Kleine Ergänzungs- bzw. Erweiterungswünsche bleiben immer offen. So wäre es beispielsweise wünschenswert, das neubenannte Kapitel Isotopenmethoden in der Biologie (in der 1. Auflage noch Tracer-Methoden in der Biologie genannt) um einige moderne Entwicklungen zu erweitern:

die cyclotron-produzierten kurzlebigen Nuklide ^{11}C, ^{13}N, ^{15}O, ^{18}F, die ^3H-NMR-Technik (in der Fußnote zu Tabelle 7.4 nur erwähnt) oder eine etwas ausführlichere Darstellung der Isotopenvariationen im natürlichen Bereich. Obwohl die $\delta^{13}C$-Definition nach (7.11) eingeführt wird, fehlt es an einem erkennbaren Bezug oder einem entsprechenden Beispiel, das diese neue Technik der natürlichen Tracer deutlich werden läßt. Da einer der Herausgeber, *H. Ziegler* — München, selbst an wichtiger Stelle zu dieser Entwicklung mit beigetragen hat, kann leicht eingeschätzt werden, zu welchen Bezügen diese Erweiterung führen kann. Skizzenhaft wäre beispielsweise folgende Konzeption denkbar: Isotopieeffekte bei der Photosynthese (Diffusion, Löslichkeit, CO_2/HCO_3^--Gleichgewicht), C_3-, C_4-Pflanzen, Atmungstests mit Mais-Glucose, $\delta^{13}C$-Werte der Isua-Sedimente auf S. 950/51, auch ein durchgerechnetes Beispiel eines Isotopieeffektes nach Gl. (7.8), Oszillationen der $\delta^{13}C$-Werte des atmosphärischen CO_2 im diurnalen und saisonalen Rhythmus als ein Beispiel für Umwelt-Biophysik, $\delta^{13}C$-Werte der *Miller*schen Synthese organischer Verbindungen aus $CO_2/NH_3/H_2O$ u. ä. m. In der Formel (7.8) sollten die ursprünglichen *Bigeleisen-Wolfsberg*-Abkürzungen (u_{1i}, u_{2i}) gewählt werden, da sonst durch die μ_A, μ'_A ... Verwechslungen mit den reduzierten Massen nach Formel (7.7) denkbar sind. Die Abb. 17.34 und die Tabelle 17.9 sollten vereinheitlicht und aktualisiert werden (Stromatolithen und nicht Stromatoliten, eine englische Version). Die Photoassimilation des CO_2 im Präkambrium durch angeblich höhere Pflanzen und Grünalgen bedarf auf S. 950 einer korrekteren Darstellung. Inwieweit die S_{AB}-Werte der 16 S ribosomalen RNA zur phylogenetischen Einordnung von Prokaryonten und Eukaryonten Erwähnung finden sollten oder die Interpretation des *Auger*-Effektes bei Biomolekülen als ein Ansatz für eine selektive Strahlenbiologie, wird immer eine Ermessensfrage der verantwortungsbewußten Herausgeber sein und bleiben.

Zur Aktualität der Literatur wäre zu sagen, daß insgesamt 826 Zitate angegeben werden, wovon allein 33,2% aus den Jahren 1977—82 stammen, also seit der 1. Auflage hinzugekommen sind. Sogar der größte Teil der Abbildungen ist mit korrekten Literaturzitaten versehen — insgesamt 411, wobei auch wieder 27,6% auf die Jahre 1977—82 entfallen.

Wissenszuwachs und ein fruchtbarer Gedankenaustausch finden heute größtenteils an den Grenzflächen von angestammten Disziplinen statt, wobei diese Quellgebiete meist durch Wortzusammenziehungen kenntlich gemacht werden; die Biophysik zählt auch dazu. Der Leser und Nutzer der vorliegenden Biophysik wird es dankbar vermerken, daß sich die Herausgeber nicht in einen sophistischen Streit über eine vielleicht definierbare wirkliche Biophysik, die zu erfassen und zu beschreiben wäre, und eine Biophysik, die mehr aus dem Wissensbestand von Nachbardisziplinen stammt (medizinische Physik, biophysikalische Chemie, physikalische Biochemie, Physiologie, molekulare Bioenergetik, Strahlenbiologie u. ä. m.), begeben haben. Natürlich bedarf die Biophysik einer sauberen Definition und klaren Abgrenzung zu anderen Disziplinen, aber dies ist kein Spezifikum dieses Faches — die chemische Physik, die Geochemie (Biogeochemie) oder etwa die Umweltwissenschaft befinden sich in gleicher oder ähnlicher Not. Durch den hier gewählten Weg wird der Einstieg zum Verständnis der komplexen Vorgänge des Lebens leichter möglich, und damit erfüllt er im vollen Umfang auch wesentliche Bedingungen, ein Lehrbuch zu sein, ein Lehrbuch der Biophysik, auf das kaum ein Student dieses Faches verzichten kann und das auch eine souveräne Übersicht für den forschend tätigen Wissenschaftler darstellt. Man sollte in künftigen Auflagen den Untertitel, ein Lehrbuch zu sein, mit Fug und Recht wieder kenntlich machen.

H. Hübner

Biogeochemistry of Ancient and Modern Environments Proc. of the 4th Internat. Symp. on Environmental Biogeochemistry and Conf. on Biogeochemistry in Relation to the Mining Industry and Environmental Pollution held in Canberra, Australia, 26 August — 4 Sept. 1979

P. A. Trudinger, M. R. Walter and B. J. Ralph (Herausgeb.). Berlin — Heidelberg — New York: Springer-Verlag 1980; 723 S., 271 Abb., 159 Tab.; DM 88,— (US $ 52,—)

In der Zeit vom 26.—31. August 1979 fand in Canberra (Australien) das 4. Internationale Symposium über Umwelt-Biogeochemie (ISEB) statt, dem in der Zeit vom 3.—4. September 1979 die Konferenz über Biogeochemie in bezug auf Bergbauindustrie und Umweltschädigung (die Leaching Konferenz) folgte. Im vorliegenden Band werden 78 Beiträge in erweiterten Fassungen wiedergegeben — rund 20% der Vorträge, die auf den beiden Konferenzen gehalten wurden, sind aus unterschiedlichen Gründen nicht enthalten.

Die zeitlich aufeinander abgestimmten Konferenztermine lassen erwarten, daß zwischen den beiden Symposien enge Wechselbeziehungen bestehen, wenn dies auch zunächst nicht augenscheinlich ist. Ein abgesteckter Rahmen, der vom Präkambrium unserer Erdentwicklung bis zu modernen Verfahren der Biometallurgie reicht, könnte als zu weit reichend angesehen werden. Wenn man allerdings nur das Beispiel der Uranium-Lagerstättenbildung herausgreift, so wird recht schnell ein bestehender Zusammenhang deutlich: Bildung der Uranium-Lagerstätten im Präkambrium und daraus sich abteilende O_2-Konzentrationen in der Atmosphäre, Anwendung der in situ-Laugung (bakterielle Laugung) von Uranium-Lagerstätten und biologische Methoden zur Entfernung von ausgewählten anorganischen Pollutanten aus Abwässern des Uranium-Bergbaues; nur ein Beispiel für die zwanglose Zusammenführung dieser Themen unter einem Aspekt, dem der Biogeochemie.

Der Präsident der Australischen Akademie der Wissenschaften, Professor *L. T. Evans*, versucht in seiner Grußadresse an die 4. ISEB die vier Wurzeln der Umwelt-Biogeochemie etwas freizulegen. Die Umweltkomponente, die in der zeitlichen Entwicklung einiger Pollutanten (CO_2, NO_x ...) lokal und global betrachtet werden kann, die biologische Komponente, die durch die mehr als 3 Milliarden Jahre wirkende Photosynthese größte Einflußnahme auf die geologische Komponente, auf die Entwicklung der Erde genommen hat, und schließlich die chemische Komponente, durch die der Mensch in größtem Umfang in die aufeinander abgestimmten Kreisläufe der Stoffe in der Natur eingreift und damit die natürlichen Ökosysteme zerstören kann.

Der Proceedings-Band gliedert sich in 3 große Abschnitte: die alte Umwelt (8 Arbeiten S. 7—92), die moderne Umwelt (38 Arbeiten S. 93—444) sowie die Biogeochemie und die Bergbauindustrie (31 Arbeiten S. 445—718). Diese großen Abschnitte werden in weitere zweckdienliche Unterabschnitte gegliedert — so in organische Geochemie oder in Schwefel-Biochemie und in Stickstoff-Biochemie. Beim Schwefelkreislauf in der Natur werden vor allem die organischen Schwefelverbindungen neben dem bekannten H_2S diskutiert, und in einer Arbeit werden Produktionsraten von Dimethylsulfid und Dimethylsulfoxid aus Meerwasser (euphotische Zone) mitgeteilt. Ebenfalls neueren Datums sind die Beobachtungen, daß N_2O nicht nur durch die bekannte Denitrifikation gebildet werden kann, sondern auch unter aeroben Bedingungen durch *Nitrosomonas* erzeugt wird, was global mit einbezogen werden muß. Sehr überzeugend sind die Beiträge über Biometallurgie, einen neuen Verfahrensweg, den man aufmerksam beachten sollte. Mit Hilfe von Mikroorganismen werden auch geringwertige Lagerstätten durch eine Anreicherung in ökonomisch vertretbare Abbauwürdigkeitsgrenzen angehoben. Die erstaunlich resistenten lithotrophen („gesteinsverzehrenden") Bakterien *Thiobacillus thiooxydans, Th. ferrooxydans* und *Ferrobacillus ferrooxydans* oxydieren minderwertige Pyrite beispielsweise zu löslichen Sulfaten, die aus den Gesteinen ausgelaugt (leaching) werden können. Durch diesen Weg könnten selbst Abraumhalden volkswirtschaftlich aufs neue interessant werden.

Zwei Beobachtungen können bei modernen Proceedings immer wieder gemacht werden: die Selbstverständlichkeit, mit der sich die Isotopenmethoden zwanglos in die methodischen und experimentellen Ergebnisse und Interpretationen einreihen, und die staunenswerte Aktualität der erfaßten Literatur. Im vorliegenden Proceedings-Band werden 1709 Arbeiten zitiert, wovon rund 48% aus den Jahren 1976–1980 stammen – die 70iger Jahre mit eingeschlossen, ergeben 80,2% der gesamten zitierten Literatur. Es lohnt sich also, diese Artikelsammlung auch von dieser Seite her zu Rate zu ziehen. Bleibt noch die Frage, wer diese Umwelt-Biogeochemie lesen sollte, wem sie zu empfehlen wäre. Von der Vielfalt der dargebotenen Themen sollte man das Buch mit einem guten französischen Menü vergleichen – so hat es auch Professer *L. T. Evans* in seiner Grußadresse getan –, es verführt dazu, von allem zu probieren. Es ist für Biochemiker, Mikrobiologen, insbesondere Geomikrobiologen, Umweltforscher und Lagerstättenkundler genauso lesenswert, wie für Isotopenchemiker und Isotopengeochemiker. Das Faktenmaterial ist umfangreich, ein Sachwortverzeichnis hilft deshalb, sich schnell zurechtzufinden und interessante Aspekte kennenzulernen – ein wertvolles und preiswertes Buch.

H. Hübner

Handbook of Material Testing Reactors and Associated Hot Laboratories in the European Community

Hrsg.: P. von der Hardt u. H. Röttger. Dordrecht – Boston – London: D. Reidel Publ. Comp. 1981. 151 S., 66 Abb., 40 Tab., 105 Lit. (engl.); geb. Dfl. 55,00; US $ 24,00

Es wird eine Übersicht über die 18 Materialtestreaktoren und großen Forschungsreaktoren der Westeuropäischen Gemeinschaft gegeben, die sich wie folgt auf die einzelnen Länder verteilen: Dänemark 1, Niederlande 1, Belgien 1, Großbritannien 2, Frankreich 6, Bundesrepublik Deutschland 6, Italien 1. Zu jedem Reaktor finden sich die wichtigsten Angaben in Form von Datenblättern: Beschreibung des Reaktors, Verwendung und Spezialisierung, Experimentelle Einrichtungen, Neutronenspektren, Charakteristik spezialisierter Bestrahlungseinrichtungen, Beschreibung der heißen Zellen, Einrichtungen für Nachbestrahlungsuntersuchungen, Verwendung der heißen Zellen.

Mit dieser kompakten und übersichtlichen Darstellung erhält der Leser einen schnellen Überblick über das Sortiment und die Experimentiermöglichkeiten der in den genannten Ländern vorhandenen Reaktoren mit einer Leistung von über 5 MW. Man erkennt auch den hohen Stellenwert, den die Kerntechnik in industriell entwickelten Ländern besitzt.

S. Niese

Handbook of Automated Analysis Continuous flow techniques

New York: Marcel Decker, Inc. 1981. 144 S.; Loseblattwerk SFr 55,00

Im vorliegenden Buch wird ausschließlich Bezug genommen auf den „Auto-Analyzer" sowie entsprechendes Zubehör der US-amerikanischen Firma Technicon Corperation. Das Grundprinzip des Auto-Analyzers ist die Analyse in einem kontinuierlich fließenden, segmentierten Flüssigkeitsstrom, in welchem spezifische chemische Reaktionen der Meßprobe mit entsprechenden Reagenzien und physikalische Prozesse (Mischung, Extraktion u. a.) stattfinden. Das System besteht aus einzelnen Bausteinen, wie Probenspeicher Dosierpumpe, Heizbad, Mischsegmenten, Dialyse-Zelle, Phasentrennung u. a., die durch Schläuche miteinander verbunden sind sowie dem Detektor und entsprechenden Anzeigegeräten, wie Schreiber oder Drucker. Neben der zur Detektion am häufigsten verwendeten Methode der Kolorimetrie werden angewandt: *p*H-Meßzellen, Spektrometrie, insbesondere UV, Fluorometrie, Atom-Absorptionsspektrometrie, Flüssigchromatographie u. a. Durch das Baukastenprinzip und die große Palette der Bausteine ist das System an viele Analysenprobleme anpaßbar. Es ist geeignet für die Untersuchung von Arzneimitteln, Pestiziden, landwirtschaftlichen Produkten, Wasser- und Luftverschmutzung, Kohlenwasserstoffen, Metallspuren in Lösungen, Farbstoffen und Pigmenten u. a. Weiterhin werden klinische Analysen sowie ozeanographische und bakteriologische Untersuchungen mit solchen Meßsystemen durchgeführt.

Das Buch ist mehr als eine Bedienungsanleitung für den Auto-Analyzer der Firma Technicon. Es vermittelt Verständnis für die Vorgänge des gesamten probenchemischen Prozesses vom Probenspeicher bis zum Detektor solcher flow-stream-Systeme, die – auch von anderen Herstellern angeboten – vielfach angewandt werden.

Es sind Richtlinien und Anweisungen zum Aufbau von flow-stream-Systemen zur analytischen Verarbeitung von wäßrigen Lösungen, organischen Lösungen, Feststoffen, Sedimenten, biologischen Geweben, Speisen und Futter, pharmazeutischen und kosmetischen Salben und Cremes angegeben.

Hinweise und Anregungen sowie praktische Tips können dem Anwender hilfreich sein, die Möglichkeiten solcher kontinuierlichen flow-stream-Systeme voll auszuschöpfen.

Der Inhalt ist übersichtlich geordnet und klar verständlich sowie gut illustriert. Das Buch gehört mit seinen praktischen Informationen in die Hand des Bedienungspersonals derartiger Analysensysteme. Eine ausführliche Darlegung der Funktionsprinzipien der wesentlichen Detektorsysteme wäre einerseits wünschenswert, würde aber andererseits den Umfang des Buches wesentlich vergrößern.

G. Meier

Heterogene Metallkomplex-Katalysatoren

G. W. Lissitschkin u. A. J. Juffa. Moskau: Verl. Chemie, 1981. 160 S., 7 Abb., 38 Tab., 432 Lit. (russ.); Broschur M 8,50

Das vorliegende Buch ist einem Gebiet der katalytischen Chemie gewidmet, das während der letzten 10 Jahre eine stürmische Entwicklung genommen hat. Es handelt sich um die Heterogenisierung katalytisch aktiver Metallkomplexe oder, wie man genauer sagen kann, um die Fixierung aktiver Metallkomplexe (vornehmlich Übergangsmetallkomplexe) an immobile organische oder anorganische Trägermaterialien. Ihre derzeit geringe Nutzanwendung in der industriellen Praxis muß hauptsächlich auf die meistens nicht einfachen Herstellungsbedingungen zurückgeführt werden.

Aus diesem Grunde legten die Autoren ihr besonderes Augenmerk auf Syntheseprinzipien solcher Katalysatoren. Nach einem recht umfangreichen Einführungskapitel zu allgemeinen Fragen der Komplexkatalyse werden im 2. Kapitel ausführlich alle bekannten Methoden der Fixierung von Übergangsmetallkomplexen auf organischen und anorganischen Trägern dargestellt. Während bei der rein adsorptiven und im wesentlichen nichtlokalisierten Fixierung von Komplexen, zu der auch die sogenannten supported liquid phase catalysts von *Rony* zu zählen sind, eine sukzessive Ablösung des aktiven Komplexes während der Reaktion in vielen Fällen, insbesondere bei Flüssigphasenreaktionen, in Kauf genommen werden muß, konnten bei der Fixierung durch Ionenaustausch durchaus stabile Systeme erhalten werden (insbesondere Arbeiten von *Haag* u. *Whitehurst*, Kap. 2.2.).

Ein besonderes Interesse beanspruchten bisher Methoden, bei denen der aktive Metallkomplex durch koordinative oder kovalente Bindung an die Trägermatrix fixiert wird. Diesen Fixierungsprinzipien wird von den Autoren im 2. Kapitel verdientermaßen breiter Raum gewidmet. Es wird gezeigt, daß durch die koordinative Bindung des aktiven Metalls an organische oder anorganische Träger mit koordinationsfähigen Ankergruppen die Möglichkeit entsteht, auch elektroneutrale Komplexverbindungen hinreichend fest an den Träger zu binden und einen modifizierenden Einfluß auf das aktive Zentrum auszuüben. Die Leistungsfähigkeit polyfunktioneller Katalysatoren, bei denen unterschiedlich katalytisch wirksame Zentren auf einer Trägermatrix angeordnet sind, ihre hohe Selektivität und Empfindlichkeit gegenüber chemischer Modifizierung wurde eindrucksvoll an den Arbeiten von *Pittman* u. Mitarb. herausgestellt. Daß auch die Trägerstruktur Einfluß auf den Charakter der Heterogenisierung und damit auf die Natur des aktiven Zentrums hat, wurde anhand der Arbeiten von *Grubbs* und *Collman* demonstriert. Hervorzuheben ist das Bemühen der Autoren, vielerorts auf offene Fragen und widersprüchliche experimentelle Tatsachen aufmerksam gemacht und kritisch kommentiert zu haben, wird doch dadurch die aktuelle Forschung auf diesem Gebiet am eindruckvollsten dokumentiert. So haben sich beispielsweise Erwartungen, die an Spacerkatalysatoren (Katalysatoren, bei denen das aktive Zentrum von der Trägeroberfläche durch eine inerte Kohlenwasserstoffkette abgesetzt ist) nicht immer experimentell bestätigen lassen (Kap. 2.3.3.). Das beweist, daß noch manche trivialen Vorstellungen durch detaillierte Untersuchungen der Zugänglichkeit und Beweglichkeit aktiver Zentren korrigiert werden müssen. Die Erfolge, die bisher mit kovalent gebundenen Übergangsmetallkomplexen an mineralischen Trägern erzielt worden sind, speziell nach ihrer reduktiven bzw. oxydativen Aktivierung, wurden insbesonders durch die Arbeiten von *Ermakov* u. Mitarb. (Katalyseinstitut Nowosibirsk) verdeutlicht.

Besondere Abschnitte im 2. Kapitel nehmen die gelimmobilisierten Komplexe von *Kabanov* und ihre Anwendung in der *Ziegler-Natta*-Katalyse, die Fixierung raumbeanspruchender Moleküle und die Heterogenisierung durch Ligandenaustausch ein.

Im 3. und letzten Kapitel gehen die Autoren auf Entwicklungstendenzen und Probleme der heterogenen Komplexkatalyse ein. Die wichtigsten Themen in diesem Zusammenhang sind die asymmetrische Katalyse mittels heterogenisierter Metallkomplexe, der Einsatz von trägerfixierten Metallkomplexen bei Hochtemperaturprozessen und die Heterogenisierung von Clusterverbindungen. Interessant ist der Vorschlag eines Klassifizierungssystems der Autoren, das durch Kombination unterschiedlicher Synthesevarianten zu neuartigen heterogenisierten Komplexkatalysatoren führen kann.

Alles in allem ein Buch, das trotz einiger Mängel und Druckfehler (z.B. falsche Tabellenangaben im Text der Seiten 74 und 79, fehlender Index bei R' S. 88, falsches Literaturzitat 176) in umfassender und übersichtlicher Weise eine neue Katalysatorklasse vorstellt, die in zunehmendem Maße an Bedeutung gewinnt. Der Wert dieser Monographie für den Fachmann wird leider dadurch eingeschränkt, daß die neuere Literatur in nur sehr geringem Maße berücksichtigt wurde. Obgleich das Buch mit dem Wissensstand von 1980 abgeschlossen wurde (Anmerkung S. 51), fand im wesentlichen nur die Literatur bis 1976 Berücksichtigung. Auf die letzten 4 Jahre von 1977—1980 entfallen lediglich 12% aller Literaturzitate.

Die Autoren haben mit ihrem Buch nicht nur in der sowjetischen, sondern auch in der internationalen Fachliteratur eine spürbare Lücke auf diesem Gebiet geschlossen (wenn man von einigen Reviewartikeln oder Spezialkapiteln in einschlägigen Büchern über Koordinationskatalyse absieht). Es kann allen denjenigen empfohlen werden, die auf dem Gebiet der Katalyse — sei es in Forschung oder Industrie — tätig sind.

H. Fuhrmann

Technische Strömungslehre
Eine Einführung in die Grundlagen und technischen Anwendungen der Strömungsmechanik

E. Becker. 5. Auflage (Teubner Studienbücher Mechanik) Stuttgart: B. G. Teubner 1982. 160 S., 134 Abb.; kart. DM 19,80

Das Lehrbuch ,,Technische Strömungslehre" befaßt sich mit den wesentlichen Grundlagen der Strömungsmechanik und deren Anwendung auf ausgewählte technische Beispiele. Der Autor hat in der gewählten Darstellung die mechanischen Inhalte strömungstechnischer Fragestellungen herausgearbeitet. Er hat ein Konzept verfolgt, das unnötige Mathematisierungen vermeidet und es gestattet, mit Hilfe der Differential- und Integralrechnung alle notwendigen mathematischen Modellierungen durchzuführen.

Im Lehrbuch werden die Eigenschaften fluider Stoffe eingeführt und dabei besonders das Fließverhalten diskutiert, wobei auch auf ,,nicht-*Newton*sche" Fluide eingegangen wird. In einem weiteren Kapitel wird die *Bernoulli*-Gleichung abgeleitet und auf einfache technische Fragestellungen angewandt. Es werden der Impulssatz und Drehimpulssatz für stationäre Strömungen behandelt, Probleme der Strömung um Flügel und Einzelgitter dargestellt, die ebene Schichtenströmung einer zähen Flüssigkeit mit vielfältigen Anwendungen auf Fragen der Viskosimetrie diskutiert. Die reibungsbehaftete Strömung durch Rohre und Rohrleitungselemente als technisch besonders wichtiger Anwendungsfall wird in einem gesonderten Kapitel abgehandelt.

Es wird auf Strömungsgrenzschichten eingegangen, und für Strömungen von Gasen werden die wichtigsten Eigenschaften und Besonderheiten gegenüber inkompressiblen Fluiden herausgearbeitet.

Der Masseerhaltungssatz (Kontinuitätsgleichung) wird im Kapitel *Bernoulli*-Gleichung eingeführt und sehr knapp dargestellt. Der Masseerhaltungssatz, der neben Energiesatz und Impulssatz zu den Grundgleichungen der Strömungsmechanik gehört, hätte seiner Bedeutung entsprechend mehr herausgearbeitet werden können.

Das Buch ist für Ingenieure und Studierende technischer Fachrichtungen geschrieben und verdeutlicht in der gewählten Darstellung besonders die Anwendung der Grundgesetze der Strömungsmechanik auf technische Problemstellungen. Die gewählten Beispiele sind sehr anschaulich und einprägsam. Auch Spezialisten aus naturwissenschaftlichen Fachrichtungen, die technisch interessiert sind, ist dieses Lehrbuch zu empfehlen.

R. Scholz

Übungen zur Technischen Strömungslehre

E. Becker, E. Pilz. 2. Aufl. (Teubner Studienbücher Mechanik) Stuttgart: B. G. Teubner 1978. 136 S., 177 Abb.; kart. DM 16,80

Dieses Buch dient der Festigung und sicheren Handhabung der strömungstechnischen Grundlagen und Hilfsmittel, die im Lehrbuch *E. Becker* „Technische Strömungslehre" hergeleitet und an wichtigen technischen Beispielen erläutert und diskutiert wurden. Die im Buch zusammengestellten 130 Aufgaben folgen inhaltlich der Einteilung im Lehrbuch und ermöglichen die Übung in der Lösung einfacher strömungstechnischer Probleme und gleichzeitig die Kontrolle in der Handhabung des erworbenen Wissens. Die Übungskapitel umfassen Aufgaben zur Hydrostatik, der Anwendung der *Bernoulli*-Gleichung, der Anwendung des Impulssatzes, den Strömungsmaschinen, dem Drallsatz, Strömungen durch Gitter, der ebenen Schichgenströmung, der Rohrströmung und der Strömung von Gasen.

Ein Beispiel der Hydrostatik ist die Gaszentrifuge zur Trennung von Isotopen. Dieses Problem wird erläutert.

Zu jedem Übungsabschnitt sind die wichtigsten Grundlagen und Prinzipien zusammengestellt, die zur Lösung benötigt werden, so daß auch eine Unabhängigkeit vom Lehrbuch erreicht wurde. Zu den Aufgaben, bei denen die Autoren Schwierigkeiten erwarten, werden zusätzliche Informationen in Form von Hinweisen und Anmerkungen, die weiterführenden Betrachtungen dienen, gegeben. Zur Selbstkontrolle sind im Lösungsteil die Ergebnisse und wichtige Zwischenschritte bei der Lösung der Übungsaufgaben aufgeführt.

Die Aufgaben beschränken sich im wesentlichen auf dichte beständige Fluide. Die Auswahl der Aufgaben und ihre Formulierung in meist technisch relevanten Sachverhalten ist den Autoren gelungen.

Die zur Lösung der Aufgaben erforderlichen mathematischen Hilfsmittel gehen nicht über die Integral- und Differentialrechnung hinaus. Das Übungsbuch ist für Studierende technischer Einrichtungen und darüber hinaus technisch interessierten Naturwissenschaftlern zur Festigung, Handhabung und Überprüfung des erworbenen Wissens auf dem Gebiet der Strömungsmechanik sehr nützlich und kann uneingeschränkt empfohlen werden.

R. Scholz

International Tokamak Reactor: Phase One Report of the International Tokamak Reactor Workshop held in seven sessions in Vienna during 1980 and 1981

IAEA, Vienna 1982, STI/PUB/619. ISBN. 92-0-131082-X; 860 S., 317 Abb., 215 Tab., 193 Lit.; kart. Austr. Sh. 1420,—

Der 1978 begründete INTOR-Workshop (International Tokamak Reactor) ist eine unter der Oberaufsicht der IAEA und speziell dem IFRC (International Fusion Research Council) stehende gemeinsame Unternehmung von Euratom, Japan, den USA und der UdSSR. Seine Hauptaufgaben sind:

1. Festlegung der Aufgaben und Charakteristika des nächsten großen Fusionsexperiments auf der Tokamak-Strecke (nach den Anlagen TFTR, JET, JT-60, T-15 und vor der Generation von Demonstrationsreaktoren)
2. Feststellung der technischen und physikalischen Grundlagen zum Aufbau eines solchen Experiments in den frühen 90er Jahren
3. Konzeption des Experiments
4. Detaillierte Ausarbeitung des Projekts
5. Aufbau und Betrieb der Anlage auf internationaler Grundlage.

Die Arbeit des Workshops geht in einzelnen Phasen vor sich. Phase Null (1979 durchgeführt) behandelte die ersten zwei angesprochenen Aufgaben. Der 1980 publizierte Bericht repräsentiert die Auffassungen der internationalen Experten und zieht den Schluß, daß ein im Zündregime mit D-T arbeitender Tokamak als ingenieurtechnische Testanlage in den frühen 90er Jahren technisch realisierbar sei, vorausgesetzt, die begleitenden Forschungen und Entwicklungen werden mit verstärkter Aktivität betrieben. Auf der Grundlage dieser positiven Bewertung entschied die IAEA, die Phase Eins des Workshops in Angriff zu nehmen, deren Aufgabe die Konzeption des Experiments war.

Der vorliegende Band, der mit einem Arbeitsaufwand von zusammen etwa 120 Mannjahren erstellt wurde, enthält diese Konzeption. In 22 Kapiteln beschreibt er die Zielstellungen der Anlage, definiert ihre Position im weltweiten Fusionsforschungsprogramm und gibt eine detaillierte technische Diskussion der Einzelfragen des Projektes. Im einzelnen geht der von den 24 Mitgliedern der INTOR-Gruppe (Euratom 9, Japan 5, USA 4, UdSSR 6) und weiteren Autoren verfaßte Bericht auf folgende Probleme ein: physikalische Grundlagen des Projektes, mechanische Konfiguration und Betriebscharakteristika, magnetisches und elektrisches System, Heizung und Brennstoffzufuhr, erste Wand und Limitersysteme, Divertoren, Tritium brütendes Blanket, Strahlenschutz, Vakuum- und Tritiumsystem, Diagnostik, Datenerfassung und Steuerung, Lageplan der Aggregate, Betrieb der Anlage und Testprogramme, Zuverlässigkeit und Zugänglichkeit der Anlage, Sicherheits- und Umweltaspekte, Kriterien zur Standortwahl, begleitende Forschung und Entwicklung, Zeitplan, Projektdaten, technische Anhänge, administrative Anhänge.

Als besonders hilfreich wird eine etwa 70 Seiten lange Zusammenfassung des gesamten Inhalts des Berichtes empfunden.

Die Ausstattung des Buches ist wie gewöhnlich bei IAEA-Publikationen ausgezeichnet. Viele Zeichnungen, graphische Darstellungen und Tabellen erlauben es auch dem Nichtexperten, sich einen Überblick zu verschaffen.

Das Buch ist unentbehrlich für direkt oder indirekt auf dem Gebiet der kontrollierten Kernfusion Tätige, dürfte aber auch für Plasmaphysiker, Kernphysiker, Kerntechniker und Energetiker von Wichtigkeit sein und darüber hinaus von allen mit Interesse aufgenommen werden, die sich über Möglichkeiten der Planung der zukünftigen Energieversorgung der Menschheit informieren wollen.

W. Hintze

Nuclear Power, the Environment and Man Information Booklet

STI/PUB/635 Vienna IAEA 1982, 195 S., 53 Schwarz-Weiß-Fotos, mehrere Farbfotos, 28 Tab.; kart. ÖS 200,—

Diese kleine Schrift ist gemeinsam von der IAEA und der WHO zusammengestellt worden und wendet sich an das breite, interessierte Publikum in Fragen der Kernenergie, ihrer Probleme, Entwicklungstendenzen und versucht vor allem die Belastungen der Menschheit durch die Kernstrahlung sachkundig darzulegen. Neben einer kurzen Darstellung der Energiebedarfsentwicklung für die vor uns liegenden Jahre und Jahrzehnte wird in anschaulicher Weise das Problem der Wiederaufbereitung von Kernbrennstoffen, der Deponie von Kernbrennstoffaschen und der Stillegung von Kernkraftwerksanlagen behandelt. Den breitesten Raum nehmen dann die Fragen der Strahlenbelastungen und des Strahlenschutzes ein, einschließlich Aussagen über radiobiologische Effekte. Sehr dienlich ist eine Zusammenstellung der meistens auftretenden Begriffe in einem Glossar, um den weniger sachkundigen Leser schnell zu informieren und zu erläutern, worum es geht. Dem sachkundigen Leser dürfte die Zusammenstellung aller wichtigen veröffentlichten Berichte, Standards, Proceedings und Technischen Reports über dieses Thema sehr dienlich sein, die sowohl von der IAEA als auch von der WHO in den letzten Jahren herausgebracht wurden.

Die Schrift ist sehr ansprechend gestaltet, die vielen Farbfotos von Kernkraftwerksanlagen und auch von einzelnen Teilansichten lockern den Stoff auf. Eigentlich sollten all diejenigen Menschen die Schrift aufmerksam lesen, die über Kernkraftwerke, Wiederaufbereitungsanlagen und Deponien diskutieren und manchmal auch Entscheidungen daraus ableiten müssen.

H. Hübner

The Study of Fast Processes and Transient Species by Electron Pulse Radiolysis

J. H. Baxendale, F. Busi. Dordrecht — Boston — London: D. Reidel Publ. Comp. 1982. 637 S., div. Abb. u. Tab.; Leinen Dfl. 175,—; US $ 74,50

Das vorliegende Buch enthält eine Sammlung von Vorträgen, die von namhaften Spezialisten im September 1981 auf einer vom NATO Advanced Study Institute in Capri, Italien, organisierten Sommerschule zum oben genannten Thema gehalten wurden.

Im ersten Hauptteil wird neben einem kurzen Abriß der Grundlagen der Absorption ionisierender Strahlung (*G. R. Freeman*, Kanada) und der chemischen Dosimetrie (*E. M. Fielden*, Großbritannien) eine detaillierte Darstellung der modernen Pulsradiolysetechnik gegeben. Ausgehend von der Charakterisierung verschiedener für die Pulsradiolyse verwendbarer Elektronenbeschleuniger (*M. C. Sauer*, USA) wird ausführlich auf alle zur Zeit bekannten Nachweistechniken kurzlebiger Zwischenprodukte (Transienten) eingegangen. Grundprinzipien, Instrumentierung, Anwendung und Leistungsfähigkeit folgender Nachweismethoden werden behandelt: optische Absorptions- und Emissionsspektroskopie (*G. Roffi*, Italien), Leitfähigkeitsmessungen in polaren und unpolaren Flüssigkeiten und polarographische Messungen (*K.-D. Asmus, E. Janata*, Berlin-West), Mikrowellenabsorptionstechnik (*J. M. Warman*, Niederlande), zeitaufgelöste und optisch nachgewiesene Elektronenspinresonanz (*A. D. Trifunac, J. P. Smith*, USA) sowie Lichtstreuung (*M. A. J. Rodgers*, USA). Die Behandlung wird ergänzt durch ein umfangreiches Kapitel über rationelle Möglichkeiten der Datensammlung, -reduktion und -analyse bei Pulsradiolyseexperimenten (*D. C. Foyt*, USA).

Im zweiten Hauptteil werden ausgewählte Ergebnisse der Anwendung der Pulsradiolyse auf anorganische, organische, mizellare und biologische Systeme dargestellt. Beginnend mit der Radiolyse des Wassers und ihrer Anwendung in der anorganischen (*G. V. Buxton*, Großbritannien), der organischen (*A. J. Swallow*, Großbritannien) und der Biochemie (*A. J. Swallow*), werden strahleninitiierte Reaktionsabläufe in Alkoholen (*G. R. Freeman*, Kanada), flüssigen Alkanen (*F. Busi*, Italien), Mizellen und anderen aggregierten Systemen (*M. A. J. Rodgers* und *B. A. Lindig*, USA) an meist gut ausgewählten Beispielen behandelt. In vier Beiträgen wird über physikalisch-chemische Phänomene wie die Dynamik von Elektronen und Ionenpaaren in unpolaren Flüssigkeiten (*J. M. Warman*, Niederlande), die Erzeugung und das Schicksal angeregter Moleküle (*M. A. J. Rodgers*, USA) und über Transienten in anorganischen und organischen Tieftemperaturgläsern (*F. Kieffer*, Frankreich) berichtet. Das Buch schließt mit einem kurzen Kapitel über die Anwendung der Pulsradiolyse zur Untersuchung von Transienten in der Gasphase (*M. C. Sauer*, USA).

Das vorliegende Buch ist keine Monographie. Es wendet sich nicht in erster Linie an auf den genannten Gebieten aktiv arbeitende Fachleute, sondern versucht zu zeigen, daß Methodik und Anwendungsmöglichkeiten der Pulsradiolyse einen Stand erreicht haben, der ihren Einsatz zur Bearbeitung einer beträchtlichen Zahl von aktuellen Fragestellungen nicht nur der Strahlenchemie, sondern auch der anorganischen, organischen und physikalischen Chemie, der Biochemie und Biologie rechtfertigt. Wissenschaftler dieser Fachrichtungen, die sich mit der Kinetik chemischer Elementarreaktionen, mit der Aufklärung von Reaktionsmechanismen, mit molekularen Transportprozessen, Energieübertragung u. ä. befassen, können von der in diesem Buch gebotenen Vielfalt des Untersuchungsmaterials profitieren.

Trotz der relativ großen Zahl von Autoren, die zu diesem Band beitrugen, ist eine befriedigende Homogenität der Darstellung erreicht worden. Zu bemängeln ist, daß die einzelnen Beiträge im Schreibmaschinensatz erschienen sind. Formeln wurden teilweise handschriftlich eingefügt. Die Sorgfalt, die bei der Auswahl der Themen von den Herausgebern *J. H. Baxendale* und *F. Busi* geübt wurde, vermißt man leider bei der Ausstattung des Buches.

R. Mehnert

Grundlagen der Kernchemie

V. Majer (unter Mitarbeit eines Autorenkollektivs). Bearb. der dt. Ausgabe: H. Kupsch. Leipzig: Verlag J. A. Barth 1982. 995 S., 208 Abb., 109 Tab.; Leinen M 110,—

Das vorliegende umfangreiche Werk ist in folgende Abschnitte gegliedert:

— Einleitung, Definition, Abgrenzungen (34 S.)

- Allgemeine Kernchemie (513 S.):
  Kernindividuen, Kernreaktionen, Chemie der gebildeten Atome, Wechselwirkung der Kernstrahlung mit Materie, Chemie der radioaktiven Elemente, Chemie der isotopen Nuklide, Chemie von Spuren radioaktiver Stoffe
- Methoden der Kernchemie (93 S.):
  Arbeitsmethoden der Radiochemie, Aktivierung, Trennprozesse, Konzentration und Isolation von Radionukliden, Trennung und Anreicherung isotoper Nuklide
- Präparative Kernchemie (84 S.):
  Herstellung natürlicher radioaktiver Stoffe, Herstellung künstlicher radioaktiver Stoffe
- Analytische Kernchemie (60 S.):
  Methoden der nuklearen Analyse, Analyse natürlicher radioaktiver Stoffe, Analyse künstlicher radioaktiver Stoffe
- Angewandte Kernchemie (145 S.):
  Allgemeine Grundlagen der Anwendung von Kernstrahlung in der Chemie, Anwendung in der allgemeinen physikalischen Chemie, Anwendung in der analytischen Chemie
- Anhang
- Literatur

Beim Lesen des Buches fällt auf, daß das Kapitel 2, „Allgemeine Kernchemie" mehr als die Hälfte des Buches ausmacht und neben Atomphysik, Kernreaktionen, Entwicklung der Radioaktivität, Kernspaltung, Strahlenchemie, physikalisch-chemischeEigenschaften der radioaktiven Elemente, Isotopieeffekte, Isotopentrennung bis hin zum Spurenverhalten radioaktiver Stoffe eine Vielzahl sehr unterschiedlicher Sachgebiete beinhaltet.

Diese Einteilungsprinzipien sind gemäß der zu Anfang getroffenen Definition und Klassifikation ungewohnt und führen zwangsläufig zu mehrfachen Wiederholungen. Im Vorwort wird zwar betont, daß solche Wiederholungen taktische Vorteile für den lernenden Studenten mit sich bringen. Dieser Meinung kann man sich aber nur bedingt anschließen, da gerade für den Anfänger die Übersicht leicht verlorengeht und ein Sachgebiet oft nicht eindeutig und in Gänze erfaßt wird. Dadurch und durch die eigenwilligen Zuordnungen und Bezeichnungsweisen ist der Text etwas umständlich und breit angelegt, dies mag z. T. auch an der Übersetzung aus dem Tschechischen ins Deutsche liegen.

Unbeschadet dessen, was in der Absicht der Autoren lag, entsprechen die einzelnen Abschnitte des Buches in Inhalt und Umfang nicht immer ihrer praktischen Bedeutung. Es werden beispielsweise allein der „Kinetik der Kernreaktionen", gemeint ist hier der Zerfall von radioaktiven Nukliden ,70 Seiten gewidmet, während die gesamte „Analytische Kernchemie" nur 60 Seiten ausmacht.

Es fällt weiterhin auf, daß die Reaktorchemie, Probleme der Wiederaufbereitung von Kernbrennstoffen, die Gewinnung seltener gering radioaktiver Elemente aus Spaltprodukten etc. praktisch gänzlich fehlen. Positiv hervorzuheben ist der Abschnitt über das Verhalten von trägerfreien Radionukliden bzw. Spurenmengen in Lösung. Dieses wichtige Gebiet findet man kaum in anderen Lehrbüchern. Insgesamt gesehen ein Lehrbuch der Kernchemie, das nicht zuletzt auf Grund des Umfanges dem Radiochemiker doch noch eine Reihe von Grundlagen ausführlich vermittelt und das auch in diesem Sinne als Nachschlagewerk verwendet werden kann. Für Studenten, insbesondere für Radiochemiker im Nebenfach, erscheint diese Ausgabe zum systematischen Lernen zu umfangreich.

Wenn auch die Meinung im Vorwort und Umschlag, daß mit diesem Buch erstmalig ein umfassendes Lehrbuch der Kernchemie in deutscher Sprache zugänglich geworden ist, zumindest seitdem es den „Lieser" gibt, nicht mehr zutreffend ist, so kann doch gesagt werden, daß dieses Buch den Radiochemikern, Fachleuten aus der Isotopen- und Strahlenforschung sowie den Studenten einschlägiger Fachrichtungen von Nutzen sein wird. Auch Chemiker und Physiker anderer Fachgebiete werden Anregungen finden können.

H. Koch

The Handbook of Environmental Chemistry
Vol. II, Part B: Reactions and Processes

O. Hutzinger (Ed.). Berlin — Heidelberg — New York: Springer 1982, 205 S., 63 Abb., 42 Tab.; Leinen DM 98,—; US $ 45,70

Dieser Teilband B des Handbuches der Umweltchemie beschäftigt sich hauptsächlich mit Reaktionen und Prozessen in der flüssigen Phase (Aquochemie) und in der Bodenzone und ist die Fortsetzung des Teilbandes A, der mehr die Vorgänge in der Troposphäre zum Gegenstand hat. Folgende 9 Themen werden im Teilband B behandelt: Umwelt-Photochemie einschließlich einiger experimenteller Aussagen über den Zugang zur Umwelt-Photochemie, aquatische Photochemie, Wechselwirkungen von Huminstoffen mit Umweltchemikalien, Komplexeigenschaften einiger Metalle mit den natürlichen oder anthropogenen Chelatbildungen, und schließlich werden Transformationen von organischen Verbindungen durch Mikroorganismen und Metabolismen in Säugetieren behandelt. Den Abschluß bildet ein kurzer Abriß über die Pharmakokinetik mit den Beispielen polychlorierte Biphenyle, polybromierte Biphenyle und dem Chloroform im Trinkwasser. An der Abfassung dieser Beiträge waren 9 namhafte Wissenschaftler beteiligt, so daß eine gut überschaubare und aus unterschiedlichen Blickrichtungen behandelte Darstellung vorliegt.

In der Darstellung der Grundprinzipien der Umwelt-Photochemie wird u. a. kurz auf die Smog-Kammer eingegangen (p. 14ff.). Diese Untersuchungstechnik einer Kinetik der Spurengase unter simulierten Bedingungen von Sonnenstrahlung, Zeitdauer, Temperatur, Luftfeuchte u. ä. Parametern ist nicht ohne Kritik in der Literatur beschrieben worden, vor allem im Zusammenhang mit dem Los Angeles Smog. Hier würde man sich wünschen, daß ein demonstratives Beispiel vorgeführt wird, einschließlich des dazu notwendigen Aufwandes an Computerexperimenten, um ungefähr die natürlichen Bedingungen gut widerspiegeln zu können. Beim Abschnitt Pharmakokinetik wird naturgemäß das Compartment-Modell als lineares Gleichungssystem zugrunde gelegt; eine Andeutung nicht-linearer Zusammenhänge (*Michaelis-Menten*-Kinetik) läßt gewisse Schwierigkeiten erahnen. Es scheint aber auch notwendig zu sein, daß ein Hinweis auf die Anwendbarkeit dieses Modells für bestimmte Ökoteilsysteme möglich ist, wie beispielsweise auf das auf S. 84 vorgestellte Modell eines Teiches, das nur durch eine Stoffbilanzgleichung beschrieben wurde.

Erstaunlich ist immer wieder die hohe Aktualität an erfaßten Literaturzitaten. Es sind in diesem Teilband 505 Literaturstellen zu finden, wobei 52,1% aus den Jahren 1975—82 stammen; selbst aus dem Erscheinungsjahr des Teilbandes sind schon 2 Zitate enthalten. Damit wird auch dieser Band nicht nur in seiner dargestellten Thematik, sondern auch durch seine weiterführende Literatur zu einem zuverlässigen Teil des Gesamtwerkes, eben eines Handbuches der Umweltchemie. Wie vielseitig heute Fragen über Vorgänge und Prozesse unserer uns umgebenden Umwelt sind, zeigt auch dieser Teilband, und allen Fragenstellenden dürfte er ein zuverlässiges Hilfsmittel im Rahmen des Gesamtanliegens des Handbuches sein.

H. Hübner

The Handbook of Environmental Chemistry
Vol. III, Part B: Anthropogenic Compounds

O. Hutzinger (Ed.). Berlin — Heidelberg — New York: Springer 1982. 210 S., 38 Abb., 53 Tab.; Leinen DM 116,—; US $ 54,00

Dieser Teilband B setzt die Reihe der chemischen Verbindungen fort, die durch den Menschen und seine Aktivitäten in die ihn umgebende Umwelt gelangen. Behandelt werden: Blei, Arsen, Beryllium, Selenium, Vanadium von den Schwermetallen und C_1/C_2-Halogenkohlenwasserstoffe (unter Ausschluß der fluorierten Verbindungen), halogenierte Aromaten, flüchtige Aromaten und schließlich oberflächenaktive organische Verbindungen. An der Abfassung dieser 6 Teilthemen waren 6 namhafte Wissenschaftler beteiligt, die den gegenwärtigen Stand über dieses wichtige Teilgebiet anthropogener Umwelteinflüsse beschrieben haben.

Die einzelnen Schwermetalle werden teilweise einschließlich der möglichen analytischen Nachweisverfahren beschrieben, bis zu recht modernen Methoden, wie der PIXE (proton-induced X-ray emission) zum Nachweis von Blei. Auf S. 40 wird zwar der analytische Nachweis von Beryllium mittels Photonenaktivierungsanalyse beschrieben, aber die Formel sollte richtig $^9Be(\gamma,n)^8B$ und $^9Be(\alpha,n)^{12}C$ heißen und nicht Y,n bzw. a,n. Vielleicht hätte man andeuten sollen, daß beispielsweise ^{124}Sb-Quellen ausreichen, um die γ,n-Reaktion auszulösen, was zur Entwicklung von tragbaren Analysengeräten (Beryllometer) geführt hat. Bei Beryllium sollte man auch nicht zwei Atomgewichtsangaben machen (S. 38), die auf der veralteten Unterscheidung einer physikalischen und einer chemischen Skala beruhen. Daß in einem solchen Handbuch auch die Geschichte der Entdeckung der behandelten Elemente mit enthalten sein sollte kann bezweifelt werden. Die vielen im Text gemachten Angaben über die Jahresproduktion einzelner Verbindungen (USA, BRD) wären besser in Tabellenform wiedergegeben worden, schon um die Trends der jährlichen Produktionssteigerung mit erfassen zu können. Die oberflächenaktiven Stoffe (Tenside), hier Surfactants genannt, werden sowohl vom Gesichtspunkt der physikalischen Chemie als auch in bezug auf Analytik auf gesetzliche Bestimmungen, auf Produktionsraten in der chemischen Industrie und auf ihr biologisches Abbauverhalten behandelt; auch hier wird ein kurzer geschichtlicher Abriß dem Beitrag vorangestellt.

Die Literaturangaben in diesem Teilband sind sehr aktuell; von den insgesamt 856 Zitaten sind allein 49,6% aus den Jahren 1976—1982. Die Hauptmenge stammt aus dem Jahre 1977 mit 106 Zitaten (entspr. 12,4%).

Dieser Teilband richtet sich betont an den Praktiker in Umweltfragen: Wie verhalten sich bestimmte Stoffe, die durch technische (industrielle) Prozesse in unsere Umwelt gelangen, und welche Folgen können sich daraus ergeben? Dabei werden allerdings keine ausführlichen pathophysiologischen oder pathobiochemischen Ausführungen gemacht, da auch manche Wechselwirkung von beispielsweise Schwermetallen mit den lebensnotwendigen Bioelementen auch heute noch nicht voll erkannt worden ist. Auch dieser Teilband — wie die anderen auch — richtet sich an Umweltforscher, Industriechemiker, Hygieniker bis hin zu Humanphysiologen, die manche Wirkung im Experiment auf ihre Ursachen hin untersuchen und hier einen Anschluß an auslösende Ursachen durch die „chemische" Umwelt erfahren.

H. Hübner

Vibrational Spectra and Structure

(A Series of Advances, Vol. 11)
Hrsg. J. R. Durig, Amsterdam: Elsevier 1982. 362 S., 110 Abb., 47 Tab.; gebunden US $ 114,00; Dfl 245,00

Das Anliegen dieser Reihe besteht darin, daß kompetente Vertreter verschiedener Arbeitsrichtungen in der Schwingungsspektroskopie anhand der neuesten Literatur und auf Grund eigener Forschungsergebnisse eine kritische Einschätzung erreichter Fortschritte und möglicher Weiterentwicklungen geben. Dabei soll zunehmend die Behandlung grundlegender Themen, die von breitem Interesse sind, im Vordergrund stehen.

Der Kenner der bisher erschienenen Fortschrittsberichte wird bestätigen, daß der vom Herausgeber erhobene Anpruch weitgehend erfüllt wird. Auch im vorliegenden Band wird die vielseitige Anwendbarkeit und übergreifende Bedeutung der Schwingungsspektroskopie deutlich, die eine Zusammenfassung von Reviews aus unterschiedlichen Fachgebieten unter dem Thema der Reihe rechtfertigt.

In Kap. 1 gibt *W. H. Weinberg* einen Einblick in die Möglichkeiten der „Inelastischen Elektron-Tunnel-Spektroskopie" zur detaillierten Untersuchung der Struktur und Wechselwirkungen von an Festkörperoberflächen adsorbierten Molekülen. Zahlreiche Anwendungsbeispiele demonstrieren die speziellen Vorzüge dieser nicht einfach zu handhabenden Technik bei der Bestimmung der katalytischen Aktivität homogener Clusterverbindungen auf Aluminium-Oberflächen.

In Kap. 2 diskutieren *Lassegues* und *Lascombe* die derzeit aus den „Schwingungsspektren gasförmiger Wasserstoffbrücken-Verbindungen" ableitbaren Informationen. Es werden dabei die Grenzen der bisherigen theoretischen und experimentellen Untersuchungsmethoden verdeutlicht und Wege zur Gewinnung genauerer Aussagen — insbesondere bei komplexen Systemen — aufgezeigt. In diesem Zusammenhang kommt auch dem Einsatz von stabilen Isotopen eine nicht zu unterschätzende Bedeutung zu.

In Kap. 3 zeigt *V. T. Aleksanyan*, welchen Beitrag die Analyse der „Schwingungsspektren von Sandwich-Komplexen" zur Entwicklung der metallorganischen Chemie in den letzten 10 Jahren geliefert hat. Entscheidende Fortschritte sind dabei erst durch die Laser-*Raman*-Spektroskopie möglich geworden. Die weitere Forschung auf diesem Gebiet wird zu neuen grundlegenden Erkenntnissen über die Natur der koordinativen Bindung führen.

In Kap. 4 geben *P. F. Lynch* und *C. W. Brown* einen Überblick über die Möglichkeiten zur „Anwendung der Schwingungsspektroskopie auf Probleme der Umweltforschung". Hochempfindliche und selektive Instrumentation sowie der Einsatz von Computern kennzeichnen die moderne Analysentechnik auf diesem Gebiet. Wesentlich Weiterentwicklungen sind hier durch Automatisierung so wie Kombination verschiedener Analysen-Methoden zu erwarten.

In Kap. 5 vergleichen *D. E. Honigs, R. M. Hammaker, W. G. Fateley* und *J. L. Koenig* die unterschiedlichen Techniken der „Zeitaufgelösten Infrarot-Interferometrie" und stellen Ergebnisse von Polymer-Streckversuchen vor, bei denen die Zeitintervalle im Mikrosekundenbereich liegen. Da es sich hierbei um eine neuere Weiterentwicklung der *Fourier*-Transform-Spektroskopie handelt, ist dieses Kap. vergleichsweise kurz und trägt mehr den Charakter einer Originalarbeit.

In Kap. 6 referiert *E. R. Bernstein* sehr ausführlich die theoretischen und experimentellen Grundlagen der „Schwingungsspektroskopie molekularer Festkörper — gegenwärtige Trends und künftige Forschungsrichtungen". Die kritische Verarbeitung der Literatur liefert zusammen mit eigenen Ergebnissen des Autors neue Einsichten in das Gebiet der Molekülkristalle.

Begrüßenswert sind die Autoren- und Sachregister am Ende des Bandes, dessen neues Format die Präsentation von mehr Information als bisher erlaubt. Der angesprochene Leserkreis ergibt sich aus den behandelten Themen, die zum überwiegenden Teil anwendungsbetont sind. Die Literaturzitate der einzelnen Reviews führen bis an das Jahr 1980 heran und sind im allgemeinen ausreichend, um einem qualifizierten Wissenschaftler die Einarbeitung in das jeweilige Anwendungsgebiet zu ermöglichen. Für den praktisch tätigen Spektroskopiker stellt das Buch eine komprimierte Information über die neuesten Fortschritte auf seinem Fachgebiet dar.

G. Hanschmann

Nuclear Medicine in Clinical Practice Selective Correlation with Ultrasound and Computerized Tomography

Ed. by L. D. Greenfield, J. M. Uszler. Deerfield Beach, Florida: Verlag Chemie International 1982. 410 S., div. Abb. u. Tab.; geb. DM 138,—

Das vorliegende Buch wendet sich nach dem Vorwort der Herausgeber an einen breiten Kreis von Medizinern, die in ihrer Zusammenarbeit mit dem Nuklearmediziner einen Überblick über diagnostische und therapeutische Verfahren der Nuklearmedizin benötigen. Es ist von einem Autorenkollektiv namhafter Vertreter der Nuklearmedizin der USA verfaßt. Von den 23 Kapiteln beschäftigen sich je zwei mit der Darstellung der elementaren Grundlagen der nuklearmedizinischen Strahlenphysik und Gerätetechnik, mit dem Strahlenschutz und der Dosimetrie, mit zu erwartenden neuen Geräten und mit potentiell neuen Radiopharmaka.

Die übrigen 19 Kapitel enthalten die klinisch angewendeten Verfahren der Nuklearmedizin. Es wird eine große Vielfalt von Methoden zur Untersuchung aller Organe und Organsysteme einschließlich der wichtigsten in-vitro-Tests geschildert. Die Durchführung, die Indikationen und Aussagemöglichkeiten sind in verständlicher Form erläutert und zum Teil in Tabellen dargestellt. Den Text ergänzen instruktive Abbildungen. Bei den szintigraphischen Organdarstellungen werden diese zum Teil kurz den durch Ultraschall und Transmissionscomputertomographie erhaltenen Ergebnissen gegenübergestellt und durch entsprechende Abbildungen dokumentiert.

Die Indikationen und die Durchführung der gebräuchlichen Therapieverfahren mit offenen Radionukliden sind ebenfalls beschrieben.

Jedes Kapitel enthält eine kurze Literaturzusammenstellung wichtiger Arbeiten zum entsprechenden Gebiet, die allerdings rein US-amerikanischen Ursprungs und teilweise schwer zugänglich sind. Eine Erläuterung spezieller Begriffe der Nuklearmedizin und ein relativ kurzes Sachwortverzeichnis ergänzen das Werk.

Durch die umfassende und gut verständliche Schilderung der nuklearmedizinischen Verfahren ist das Buch als Nachschlagewerk zur raschen Information für alle klinischen Disziplinen wertvoll, besonders aber für den in der Nuklearmedizin tätigen Arzt.

M. Krüger

Lucy Die Anfänge der Menschheit

D. Johanson u. M. Edey. München — Zürich: R. Piper Verlag 1982. 492 S., 32 Schwarz-Weiß-Fotos u. 142 Strichzeichn., Diagr. u. Tab.; geb. DM 38,—

Im 17. Jahrhundert lebte in Düsseldorf der Lehrer *Joachim Neumann.* Der Mode der Zeit folgend, gräzisierte er seinen Namen und nannte sich fortan *Neander.* Er liebte die Natur, schrieb Gedichte und komponierte Kirchenlieder. Besonders schätzte er ein kleines Tal der Düssel, und um ihn zu ehren, nannten die Einwohner Düsseldorfs dieses Tal später das Neandertal. Daß dieser Name und dieses Tal einmal so berühmt werden sollten, war kaum zu erwarten, doch heute hat jeder eine Assoziation oder eine vage Vorstellung, die mit diesem Namen verknüpft ist. Als im August 1856 in einer Grotte eben dieses Neandertales Knochen gefunden wurden und ihre Zuordnung so gar nicht gelingen wollte, da waren weder Darwins revolutionierende Ideen veröffentlicht, noch hatte man sich Vorstellungen darüber gemacht, wie stark sich eigentlich menschliche Knochen im Evolutionsprozeß verändern können. Was nimmt es wunder, wenn von diesen Knochenfunden solche Annahmen gemacht werden konnten wie: Schädel eines Kosaken, der sich 1814 bei der Verfolgung der napoleonischen Armee in eben diese Höhle verirrt habe, oder Schädel eines Kelten, der in mancher Hinsicht einem heutigen Iren gleiche; selbst der berühmte *Rudolf Virchow* nahm an, daß es sich um pathologische Veränderungen handele, die mit urgeschichtlichen Formen einer Menschwerdung nichts zu tun hätten. Er schreibt u. a., daß die Mißbildungen Folge einer Rachitis im Kindesalter seien und der Schädel des Erwachsenen durch schwere Schläge deformiert worden wäre.

Welch ein Wandel seither in den Ansichten über Fossilien, und mit welcher Selbstverständlichkeit werden Fossilien heute als Zeugen und Beweise einer Evolution angesehen. Die Wissensdisziplin über die Entwicklung des Menschen, die Paläoanthropologie, ist durch aufsehenerregende Funde der letzten Jahrzehnte, ja der letzten Jahre, zu neuen Einsichten vorgestoßen, die anderen Entdeckungen an Erstaunlichem kaum nachstehen. Eine dieser jüngsten Entdeckungen ist Lucy, ein Fund hominider Fossilien im Afar-Dreieck (Äthiopien) am 30. November 1974 durch den Autor des Buches.

Lucy war ein weibliches aufrechtgehendes Wesen, nur 1,07 m groß, und lebte vor über 3 Millionen Jahren. Die Namensgebung ist auf einen Beatles-Song (Lucy in the Sky with Diamonds) zurückzuführen, der von den Expeditionsteilnehmern damals vor lauter Freude über den gemachten Fund pausenlos und lautstark vom Tonband abgespielt wurde.

Das Buch besteht, vereinfachend gesagt, aus zwei großen Teilen: einem ersten Teil, der die Geschichte von Fossilienfunden insgesamt beschreibt, und einem zweiten Teil, der solche Fragen zu beantworten versucht wie: Was ist Lucy? und Warum ging Lucy aufrecht? Überlagert werden beide

Teile durch viele interessante Einzelheiten, persönliche Eindrücke und Erlebnisse und durch wiedergegebene Gespräche mit bekannten und berühmten Paläoanthropologen. Naturgemäß nehmen Probleme der Datierung einen breiten Raum ein, und es ist ein Schulbeispiel dafür, wie die K-Ar-, die Fission-Track- und die Paläomagnetismus-Methode mit den (bio)-stratigraphischen Methoden gekoppelt werden, um verbindliche Aussagen machen zu können, einschließlich der F-Methode aus den 40er Jahren. Die Geschichte der K-Ar-Datierungen an jüngeren Gesteinsproben wird etwas ungenau wiedergegeben. Auf einem Jahrestreffen der Pazifischen Sektion der Petroleum-Geologen in Los Angeles werden schon 1956 Datierungen von Proben vorgestellt, die ein Alter von kleiner als 2 Millionen Jahren hatten. Die Frage bleibt natürlich, ob ein Paläoantropologe auch Tagungen von Erdölgeologen besuchen soll oder zumindest ihre Literatur lesen muß. Auf S. 232 wird gesagt, daß das Massenspektrometer von *James Aronson* „nur einige Atome" messen könne, wohl kaum. Die kleine Überschlagsrechnung auf S. 238 über die Anzahl der Ar-Atome ist durch die Kommastellenangabe falsch. Auf S. 246 wird der heutige normale Zustand des Erdmagnetfeldes mit dem magnetischen Nordpol in der Nähe des geographischen Nordpols angegeben, mitnichten. In der paläomagnetischen Geschichte auf S. 249 sind nur die langzeitigen Umpolungen ($5 \cdot 10^5$ bis 10^7 a) aufgeführt, nicht aber auch alle kurzzeitigen Umpolungen (Ereignisse von 10^4 a), die diesem Datierungsverfahren seine Schwierigkeiten aufprägen. Außerdem heißt es Cochiti- und nicht Cochitit-Ereignis. Auf den S. 448/49 ist dreimal von Silikon die Rede, obwohl Silizium gemeint ist, Silizium in den Zähnen; ein Übersetzungsfehler.

Der zweite Teil, der die vielen Fragen nach Herkunft und Einordnung von Lucy zu beantworten versucht, ist vor allem durch die Argumentationskette des Bewegungsanatomen *C. Owen Lovejoy* von der Kentucky State University geprägt, der u. a. die K-Strategie zur Erhaltung der Art insbesondere bei den Primaten als einen wesentlichen Faktor zur Weiterentwicklung des Gehirns und damit der Hominiden ansieht. Daß diese einfache Ausgangsstrategie nicht ausreicht, um den Zeitpunkt der Divergenz von Affe und Mensch zu bestimmen, muß zu manchen emotionalisierten Gegenargumenten führen, obwohl die Kenianischen Fußspuren von Laetoli ($3{,}75 \cdot 10^6$ a) gut die noch frühere Trennung andeuten. Außerdem hat der Autor bereits 1973 ein Kniegelenk (Hadar-Region) gefunden, das auf etwa $4 \cdot 10^6$ a datiert werden konnte und als ein Indiz für einen aufrechten Gang angesehen wird. In allerjüngster Zeit sind weitere Funde bekannt geworden, die um eine weitere halbe Million Jahre älter sein sollen als Lucy. Welch eine Zeitspanne: Wenn man den eingangs erwähnten Neandertaler in einen Zeitraum von 34000 ... 100000 Jahren zurückdatiert, also 1200 ... 3000 Generationen vor uns leben läßt, dann sind bei Lucy noch weitere 100000 Generationen hinzuzuzählen, um in die Nähe dieser vergangenen Zeiten zu kommen. Hier scheinen derzeit die Vorstellungen über den Evolutionsdruck und die Evolutionsgeschwindigkeit zur Weiterentwicklung noch in der Diskussion der Gelehrten zu sein, obwohl die Autoren ihre Auffassung überzeugend und attraktiv darzustellen verstehen, stellenweise sogar bis ans Sensationelle reichend.

Das Buch ist fesselnd und vorzüglich geschrieben; Für alle, die etwas über die Entwicklungsgeschichte des Menschen wissen möchten, ist es ein wertvolles und informatives Buch; es erübrigt sich fast, ihm eine besondere Empfehlung mit auf den Weg zu geben. Dem Physiker, der sich mit der Anwendung und Pflege von Datierungsmethoden für das späte Tertiär beschäftigt, dient es als Anregung zur Suche neuer Wege und als schöner Beweis seiner Methoden auf einem so zentralen Thema: den Menschen betreffend.

H. Hübner

Elementarvorgänge fluider Wirbelbewegungen

Von WERNER ALBRING

1981. X, 305 Seiten — 95 Abbildungen — 114 Tafeln — gr. 8° — Leinen 38,— M;
Ausland 48,— M
Bestell-Nr. 762 795 1 (6574)

Das Buch bietet Ergebnisse strömungsmechanischer Forschung aus einem Gebiet, dem in der Gegenwart besonderes Interesse gilt, das aber in der Vergangenheit nur wenig beachtet wurde, es berichtet über das Bewegen von Einzelwirbeln, von ganzen Wirbelfeldern und von Wirbelfeldsystemen in ihrer Abhängigkeit von den Koordinaten des Weges und der Zeit. Während langer Arbeitsperioden ist von den Strömungsmechanikern bevorzugt der Weg verfolgt worden, für verwirbeltes Fließen die Zeitabhängigkeit durch statistisches Mitteln zu beseitigen. Aber um einen Ablauf soweit zu verstehen, daß er durch kleine Steuerimpulse in einem gewünschten Sinne beeinflußbar werden kann, muß das Kausalgeschehen durchschaut und muß der elementare Vorgang auch in seiner Zeitabhängigkeit studiert werden.
Das Buch entstand an einer Technischen Universität, es stützt sich auf fünfzehnjährige kollektive Forschungsarbeit und auf didaktische Erfahrungen über das Darlegen von Ergebnissen in den Vorlesungen.

Bitte richten Sie Ihre Bestellungen an eine Buchhandlung

THEORIA CUM PRAXI

AKADEMIE-VERLAG
DDR-1086 Berlin, Leipziger Str. 3—4

Contents of this Issue

Articles

Letters to the Editor

Indexed in Current Contents/PC&ES
Scientific Citation Index
and INIS Atomindex

FSC
www.fsc.org
MIX
Papier | Fördert
gute Waldnutzung
FSC® C083411

Zeitfracht Medien GmbH
Ferdinand-Jühlke-Straße 7
99095 Erfurt, Deutschland
produktsicherheit@kolibri360.de